Preface

It is now 17 years since the junior author's book *Parasitic protozoa* was first published, and 13 years since it received limited revision. The study of symbiotic protozoa has meanwhile progressed, and much of the content of the earlier book has been superseded if not displaced by recent knowledge. We believe that there is still a place for an introductory textbook, conventionally arranged on a taxonomic framework, on this most fascinating group of organisms. It should supplement (certainly not supplant) more modern treatments from different angles. A mountain looks very different if one approaches it from different sides, though a true picture of that mountain can be built up only by the laborious summation of the views provided by all approaches.

The immunology and the chemotherapy of protozoal infections are subjects so complex that we do not propose to try to encompass them in this book. We mention briefly the more commonly used therapeutic compounds in the various chapters that follow, without giving details of dosage schedules. Further information on immunology and chemotherapy can be obtained from the books listed at the end of the Introduction; the information we give on chemotherapy is largely drawn from that of James & Gilles.

This book can be regarded as an evolutionary descendant of the earlier "Parasitic protozoa", referred to above, extensively revised and much more fully illustrated. We hope that it will prove as useful as its predecessor appears to have been.

J. P. Kreier
J. R. Baker

v

Acknowledgements

In the preparation of this book we have benefited greatly from discussions with many colleagues. We particularly thank Professor F. E. G. Cox (University of London, King's College;) for reading the entire typescript and making many valuable suggestions. The responsibility for any remaining errors, omissions or idiosyncrasies is, of course, ours.

Many of the photomicrographs not otherwise attributed are from a collection given to the senior author by the late Wolfram Kretchmar. Other illustrative material has been provided by colleagues or reproduced, by permission, from other publications, as listed below.

Fig. 2.1: K. Vickerman (University of Glasgow) & F. E. G. Cox (University of London), from *The protozoa* (1967, John Murray). Figs 2.3, 2.9: Xu Linhe (Shanghai Medical University). Figs 2.5, 2.12, 2.20B, 8.2: Georg Thieme Verlag (from J. P. Kreier *et al.*, 1975, *Tropenmedizin und Parasitologie*, **26**, 9–18); also Fig. 2.10C (from J. R. Seed *et al.*, 1973, *loc. cit.*, **24**, 146–60) and Fig. 3.78 (from J. R. Seed *et al.*, 1973, *loc. cit.*, **24**, 525–35). Figs 2.6, 2.8, 2.17: M. Aikawa (Case Western Reserve University). Fig. 2.15: T. Byers (Ohio State University). Fig. 2.16: Springer-Verlag (from E. Scholtyseck, 1979, *Fine structure of parasitic protozoa*). Fig. 2.21: Academic Press (micrograph provided by H. G. Sheffield and published in J. P. Dubey, 1977, in *Parasitic protozoa*, J. P. Kreier, ed., **3**, 101–237); also Fig. 9.4 (from E. U. Canning, 1977, *loc. cit.*, **4**, 195–6), Fig. 2.20A (J. R. Seed *et al.*, 1971, *Experimental parasitology* **30**, 73–81), Fig. 3.7 (from J. R. Seed *et al.*, 1972, *loc. cit.*, **31**, 399–406). Figs 8.4, 8.7 (from H. Mehlhorn & E. Schein, 1984, *Advances in Parasitology* **23**, 37–103); Fig 3.1: P. T. K. Woo (University of Guelph). Fig. 3.16: *Canadian Journal of Comparative Medicine* (from P. T. K. Woo *et al.*, 1970, vol. **34**, 142–7). Fig. 3.4: London School of Hygiene & Tropical Medicine (from J. J. Shaw, 1969, Memoir no. 13). Fig. 4.8: D. Bermudes (Boston University). Fig. 6.1: St Martin's Press (from R. D. Manwell, 1961, *Introduction to protozoology*, after Iitsuka). Fig. 6.16: American Veterinary Medical Association (from J. P. Dubey), 1976, *Journal of the American Veterinary Medical Association* **169**, 1061–78). Fig. 8.5: O. P. Gautam (Haryana Agricultural University). Fig. 9.1: L. G. Mitchell (Iowa State University). Figs 9.2, 9.5: J. Shadduck (University of Illinois). Fig. 9.3: Charles C. Thomas, publishers (from R. R. Kudo, 1966, *Protozoology*, edn 5). Fig. 10.4: B. A. Dehority (Ohio Agricultural Research & Development Center). Table 6.1 is based on material given by L. P. Pellérdy (1965, *Coccidia and coccidiosis*: Budapest, Akademiai Kiado) and Table 6.4 is based on material given by N. D. Levine & W. Tadros (1980, *Systematic Parasitology* **2**, 41–59).

PARASITIC PROTOZOA

J. P. KREIER

Department of Microbiology, Ohio State University

AND

J. R. BAKER

*The Culture Centre of Algae and Protozoa,
Institute of Terrestrial Ecology,
Cambridge, England*

Boston
ALLEN & UNWIN
Wellington   London   Sydney

Allen & Unwin, Inc.,
8 Winchester Place, Winchester, Mass. 01890, USA

Allen & Unwin (Publishers) Ltd,
40 Museum Street, London WC1A 1LU, UK

Allen & Unwin (Publishers) Ltd,
Park Lane, Hemel Hempstead, Herts HP2 4TE, UK

Allen & Unwin (Australia) Ltd,
8 Napier Street, North Sydney, NSW 2060, Australia

First published in 1987

British Library Cataloguing in Publication Data

Kreier, Julius P.
   Parasitic protozoa.
1. Veterinary protozoology   2. Protozoa,
Pathogenic
I. Title   II. Baker, John R.
593.1′045249   SF780.6
ISBN 0–04–591021–9
ISBN 0–04–591022–7 Pbk

Library of Congress Cataloging-in-Publication Data

Kreier, Julius P.
   Parasitic protozoa.
Includes bibliographies and index.
1. Protozoa, Pathogenic. I. Baker, John R.
(John Robin) II. Title. [DNLM: 1. Parasites.
2. Protozoa. QX 50 K92p]
QR251.K74   1987   593.1′045249   87–1486
ISBN 0–04–591021–9 (alk. paper)
ISBN 0–04–591022–7 (pbk. : alk. paper)

Set in 10 on 12 point Palatino by Columns, Caversham, Reading and
printed in Great Britain by St Edmundsbury Press Limited
Bury St Edmunds, Suffolk.

Contents

CONTENTS

List of tables

Introduction

In the first chapter of this book an attempt is made to answer the question "What is a protozoon?" In this introduction it is necessary to consider briefly the related question "What is a parasite?", and to define some terms.

The phenomenon of parasitism will be considered here only very briefly. The *Concise Oxford English Dictionary* (fourth edition 1949) states that a parasite is an "animal . . . living in or upon another and drawing nutriment directly from it", and derives the word "parasite" from the Greek *para* ("beside, beyond, wrong, irregular") and *sitos* ("food").

Webster's New Collegiate Dictionary (second edition 1949) states that parasitism is an antagonistic symbiosis. **Symbiosis** (Greek *syn-* or *sym-*, "together", and *bios*, "life") is defined in the same dictionary as "the living together in intimate association or close union of two dissimilar organisms". Symbiosis in this usage is said to include parasitism or antagonistic symbiosis. In zoology, **parasitism** may be broadly defined as an association between two animals of such a kind that one lives and feeds, temporarily or permanently, either in or on the body of the other. This definition includes the mammalian fetus, and it is probably fair to regard this association as an example of a rare phenomenon, **intra-specific parasitism** (also exemplified by the ecto-parasitic males of certain Crustacea (barnacles) and deep-sea angler fish). Parasitism between animals belonging to different species can be referred to as **inter-specific parasitism**. The use of the word "feeds", in the above definition, is deliberate: an association between two animals in which one is merely transported from place to place on the body of the other, without feeding while thereon, is called **phoresy** (from the Greek *phero*, "to bear").

In any symbiotic association including parasitism the smaller of the associating pair of animals is regarded as the **symbiont** or **parasite** and the larger as the host. According to whether the symbiont or parasite lives in or on its host's body, the association is referred to as **endo-** or **ectosymbiosis** or parasitism respectively. Some symbionts, or some stages in the life cycle, are unable to life apart from their hosts: they are called **obligate** symbionts or parasites. Others can live equally well as symbionts or free: these are **facultative** symbionts.

Sprent differentiates between true parasitism and commensalism. In Sprent's system **commensalism** (Latin *com-*, "together", and *mensa*, "a table") is an association in which the commensal does not feed on the host's tissues. Both ecto- and endocommensalism occur. Certain ciliates live as ectocommensals on the body surfaces of fish. Most intestinal parasites of man and other animals are really endocommensals. Their hosts are only marginally affected, if at all, by the fact that the parasite diverts some of the host's food to its own use. In Sprent's system true parasitism is, by contrast, an association in which the parasite feeds on the tissues of the host and thereby does it harm. Some people restrict the use of the word parasitism to this particular sense; it may perhaps be differentiated as **tissue parasitism**.

Other workers define parasitism as an association between two organisms "in macromolecular contact" with each other. This implies that the parasite takes in macromolecules (proteins, etc.) from the host and releases others into the host. This is really another way of defining "tissue parasitism". Yet another definition is the "genetic" one. By this definition a parasite is an organism which is dependent upon the host for a minimum of one gene or its product. Not all tissue parasites are necessarily, or always, harmful to their hosts: many trypanosomes (e.g. *Trypanosoma lewisi*) are undoubtedly tissue parasites as they feed on blood plasma but apparently do no harm to their hosts; others may be exceedingly harmful, "even unto death" (e.g. *T. brucei gambiense* in man). Such harmful parasites are said to be **pathogenic**. Some symbionts in the broad sense live sometimes as commensals and at other times as tissue parasites (e.g. *Entamoeba histolytica* in man).

Mutualism is the name given to associations in which both partners benefit. Proven examples are rare among protozoa, but a classic one is the association between hypermastigid flagellates and termites, in which the flagellates, unable to survive outside the gut of their host, digest the cellulose on which the termite feeds and which the termite itself is entirely unable to digest; thus without its flagellates, the termite dies of starvation no matter how abundant its food supply. The relationship between the fungal and algal constituents of lichens is another classic example of mutualism. Mutualistic protozoa, as far as is known, always live within the bodies of their hosts, but mutualistic relationships involving one organism living on the body surface of its host are known among other phyla. For example, sea anemones living on the shells of hermit crabs, the former benefiting from the provision of transport and particles of food dropped by the crab and the latter being defended against predators by the cnidoblasts of their passengers.

Thus a variety of definitions and usages of the terms parasitism and symbiosis occurs in the literature. Parasitism is sometimes used as the broad term meaning living together with no implication of harm or benefit to either member of the pair. In this usage mutualism is a subdivision of parasitism. Sometimes symbiosis is considered synonymous with mutualism. In all these systems the common view that parasites are undesirable to their hosts is obscured and no single word to describe the association which is detrimental to the host is available.

We will use symbiosis in this book as the general term meaning living together. It includes parasitism, a symbiotic relationship in which one member (the parasite) benefits and the other (the host) may be harmed, commensalism in which one member (the commensal) benefits and the other member (the host) is not harmed, and mutualism, in which both members benefit. As is often true in biology, the distinction between these different states is not always clear-cut, especially as many protozoa may often live as harmless commensals within their hosts, becoming harmful parasites only occasionally, for reasons which are seldom understood; such organisms we regard as parasites, as they have the potential to harm their hosts. However, because the main emphasis of this book is on the parasitic forms, and also for the sake of convenience, we have retained the adjective "parasitic" in the book's title.

Many symbionts (**monoxenous** forms) have only a single host during their life cycle, part of the latter being spent outside the host; others (**heteroxenous** forms) have two, occasionally more, hosts, usually belonging to widely separated taxonomic groups. The two hosts are sometimes distinguished as the **definitive host** (in which the symbiont undergoes sexual reproduction) and the **intermediate host** (in which it does not). This distinction is, however, impossible with organisms like the trypanosomes, in which a sexual process, if it occurs, has not yet been described, and inappropriate with those like the malaria parasites, in which the "intermediate" host (the vertebrate) is probably the one in which the protozoon's ancestors evolved, the second "definitive" host (an insect) being adopted only relatively late in the evolutionary process, and in which the sexual cycle starts in one host and is completed in the other. Thus, at least in protozoology, these terms are seldom used. One of the hosts is, instead, often referred to as the **vector**, a term which is difficult to define objectively, but which implies that the host so named transmits the protozoon to the other host. From the human point of view the mosquito is the vector of malaria; but from the mosquito's viewpoint, man is the vector. In practice, the term vector is restricted to the invertebrate host when the other host is a vertebrate or a plant. If the protozoon grows and reproduces in the vector, its transmission is said to be **cyclical**; the case where no growth or reproduction occurs in the vector is referred to as noncyclical or "mechanical" transmission.

FURTHER READING

General

Baker, J. R. 1982. *The biology of parasitic protozoa*, Studies in biology 138. London: Edward Arnold.

Cheng, T. C. 1976. *General parasitology*. 2nd edn. New York: Academic Press.

Cohen, S. & K. S. Warren (eds) 1982. *Immunology of parasitic infections*, 2nd edn. Oxford: Blackwell Scientific Publications.

Cox, F. E. G. (ed.) 1982. *Modern parasitology*. Oxford: Blackwell Scientific Publications.

Dogiel, V. A. 1964. *General parasitology*, revised and enlarged by Y. I. Polyanski & E. M. Kheisin (translated by Z. Kabata). Edinburgh: Oliver & Boyd.

Kreier, J. P. (ed.) 1977. *Parasitic protozoa*, 4 vols. New York: Academic Press.

Lee, J. F. A., S. H. Hutner & E. C. Bovee (eds) 1985. *An illustrated guide to the protozoa*. Lawrence, Kansas: Society of Protozoologists.

Schmidt, G. D. & L. S. Roberts 1985. *Foundations of parasitology*, 3rd edn. St Louis: Times Mirror/Mosby.

Genetics

Walliker, D. 1983. *The contribution of genetics to the study of parasitic protozoa*. Letchworth, England: Research Studies Press, and New York: Wiley.

Chemotherapy

James, D. H. & H. M. Gilles 1985. *Human antiparasitic drugs: pharmacology and usage*. Chichester: Wiley.

Schoenfield, H. (ed.) 1981. *Antiparasitic chemotherapy*. Basel: Karger.

Immunology

Kreier, J. P. (ed.) *Malaria*. Vol. 3: *Immunology and immunization*. New York: Academic Press.

Porter, R. and J. Knight (eds) 1974. *Parasites in the immunized host: mechanisms of survival*. Ciba Foundation Symposium 25. Amsterdam: Associated Scientific.

Royal Society 1984. *Towards the immunological control of human protozoal diseases*. London: Royal Society.

Tizard, I. (ed.) 1985. *Immunology and pathogenesis of trypanosomiasis*. Boca Raton, Florida: CRC Press.

Wakelin, D. 1984. *Immunity to parasites*. London: Edward Arnold.

CHAPTER ONE

Classification and evolution of the symbiotic protozoa

CLASSIFICATION

The protozoa may be classified as a phylum within the animal kingdom or they may, probably more sensibly, be included in the kingdom Protista. If the latter approach is taken the major groups of protozoa become phyla. The protozoa are eukaryotic organisms (eukaryotes) – those in which, *inter alia*, the deoxyribonucleic acid (DNA) is contained within a membrane-bound nucleus. They are contrasted with prokaryotes (bacteria and cyanophytes or "blue-green algae") in which this is not so. Protozoa include the smallest eukaryotes (though the largest protozoa are bigger than the smallest metazoa) and are often said to be the most primitive. This is true inasmuch as they probably differ the least from the original hypothetical group of living organisms that was ancestral to the members of both the plant and animal kingdoms, but protozoa have, of course, been evolving for just as long as we have, and some of them are very highly evolved.

The fact that the protozoa are the modern representatives of the group which was ancestral to both animals and plants leads to taxonomic difficulties. Among the Mastigophora (flagellates), some forms possess chloroplasts and have typically plant-like (holophytic) nutrition; others, obviously closely related, lack chlorophyll and have typically animal-like (holozoic) nutrition, and some can function either way depending on circumstances. The suggestion made by Haeckel in 1866 that the two groups should not be separated but included with all the other protozoa and eukaryotic unicellular algae in the kingdom Protista is basically sound, but only recently has it begun to be generally adopted. Slow change in taxonomy is not altogether bad, however, as taxons are like words in a language and rapid change in

1

taxons, like rapid change in the meanings of words, would interfere with mutual understanding.

The Protista may be regarded as a group of eukaryotes which have not adopted multicellular somatic organization, involving the development of tissues. This circumlocution avoids the necessity of having to state categorically that the Protista are either unicellular or noncellular. Both concepts are true; it depends on the angle from which one views the question. Structurally the Protista are undoubtedly unicellular organisms: the anatomy of a protistan is basically identical with that of a single metazoan cell. Functionally, however, a protistan is undoubtedly noncellular in the sense that it is a whole complete organism – just as much as an elephant or a tree is – but it has not adopted the expedient of dividing itself into a large number of small structural units or cells. One result of this has been to limit the size of Protista: and this has limited the extent to which Protista have been able to become independent of their environment. In this sense, they are primitive. In terms of the number of individuals alive at any given time they are probably the most successful of all eukaryotes.

As noted above, the photosynthetic and heterotrophic Protista are closely interrelated; this said, we shall now arbitrarily ignore all those Protista which contain chlorophyll; the residue we shall loosely refer to as protozoa (with a lower-case initial). Not surprisingly, there is not complete agreement as to how the protozoa should be classified. It is perhaps important to remember that taxonomists are often attempting the impossible: trying to impose arbitrary division on that which is in fact a continuum and thus criticism is unfair. The scheme proposed by a committee of the Society of Protozoologists in 1980, in general, is used in this book, as set out below. The main diagnostic features of each group are given and the names of symbiotic genera discussed in the book are included (in italics).

The separation of the kingdom of Protista into phyla, as has been adopted by J. O. Corliss (1984) and F. E. G. Cox (1981) and in this book, is the latest step in the evolution of protistan taxonomy. It is undoubtedly the correct approach, but will require much time and study to be developed and generally accepted.

Kingdom PROTISTA Haeckel, 1866

Phylum SARCOMASTIGOPHORA Honigberg and Balamuth, 1963. Single type of nucleus, except in some Foraminiferida; sexuality, when present, essentially syngamy; flagella, psuedopodia or both.

Subphylum MASTIGOPHORA Diesing, 1866. One or more flagella typically present in trophozoites; asexual reproduction basically by intrakinetal (symmetrogenic) binary fission; sexual reproduction known in some.

2

Class PHYTOMASTIGOPHOREA Calkins, 1909. Typically with chloroplasts or relationship to pigmented forms containing chlorophyll clearly evident; mostly free-living.

Class ZOOMASTIGOPHOREA Calkins, 1909. Chloroplasts absent; one to many flagella; ameboid forms, with or without flagella, in some groups; sexuality known in few groups.

Order KINETOPLASTIDA Honigberg, 1963, emend. Vickerman, 1976. One or two flagella typically with paraxial rod in addition to axoneme; single large mitochondrion (sometimes nonfuctional), normally containing one or more conspicuous aggregations of DNA (kinetoplast or nucleoid) located near flagellar kinetosomes; mostly symbiotic.

Suborder BODONINA Hollande, 1952, emend. Vickerman, 1976. Typically two dissimilar flagella; kinetoplast DNA aggregated in single mass, often large (eukinetoplastic condition), or in several discrete bodies (polykinetoplastic condition), or dispersed throughout mitochrondrion (pankinetoplastic condition); free-living or symbiotic. *Cryptobia*.

Suborder TRYPANOSOMATINA Kent, 1880. Single flagellum either free or attached to body by undulating membrane; kinetoplast single, relatively small and compact; symbiotic. *Blastocrithidia, Crithidia, Endotrypanum, Herpetosoma, Leptomonas, Leishmania, Phytomonas, Rhynchoidomonas, Trypanosoma*.

Order RETORTAMONADIDA Grassé, 1952. Two to four flagella, one turned posteriorly; symbiotic. *Chilomastix, Retortamonas*.

Order DIPLOMONADIDA Wenyon, 1926, emend. Brugerolle, 1975. One or two symmetrically arranged karyomastigonts (association of nucleus with one to four flagella), typically one flagellum recurrent; free-living or symbiotic. *Enteromonas, Giardia, Hexamita, Octomitus, Spironucleus*.

Order TRICHOMONADIDA Kirby, 1947, emend. Honigberg, 1974. Typically karyomastigonts with four to six flagella, one recurrent, sometimes with undulating membrane; all or nearly all symbiotic. *Dientamoeba, Histomonas, Trichomonas*.

Order HYPERMASTIGIDA Grassi and Foa, 1911. Numerous flagella arranged in complete or partial circle, in plate or plates, or in longitudinal or spiral rows meeting in a centralized structure; one nucleus per cell; all symbiotic in termites, woodroaches or cockroaches ("termite flagellates"). *Trichonympha*.

Subphylum OPALINATA Corliss and Balamuth, 1963. Numerous cilia in oblique rows over entire body surface; cytostome absent; binary fission generally interkinetal (symmetrogenic); known life cycles involve syngamy with anisogamous flagellated gametes; all symbiotic.

Class OPALINATEA Wenyon, 1926.

Order OPALINIDA Poche, 1913. *Opalina*.

Subphylum SARCODINA Schmarda, 1871. Pseudopodia, or locomotive protoplasmic flow without discrete pseudopodia; flagella, when present, usually restricted to developmental or other temporary stages; body naked

3

or with external or internal test or skeleton; asexual reproduction by fission; sexuality, if present, associated with flagellate or, more rarely, ameboid gametes; most species free-living.

Superclass RHIZOPODA von Siebold, 1845. Locomotion by lobopodia, filopodia, or reticulopodia, or by protoplasmic flow.

Class LOBOSEA Carpenter, 1861. Pseudopodia lobose, or more or less filiform and produced from broader hyaline lobe; usually uninucleate.

Subclass GYMNAMOEBIA Haeckel, 1862. Naked (without test or shell).

Order AMOEBIDA Ehrenberg, 1830. Typically uninucleate; no flagellate stage. *Acanthamoeba, Endolimax, Entamoeba, Iodamoeba.*

Order SCHIZOPYRENIDA Singh, 1952. Temporary flagellate stage in most species. *Naegleria.*

Phylum LABYRINTHOMORPHA Page, 1980. No symbiotic forms

Phylum APICOMPLEXA Levine, 1970. Apical complex, visible with electron microscope, present at some stage; micropore(s) generally present at some stage; cilia absent, flagella on sexual stages only; sexuality by syngamy; all symbiotic.

Class SPOROZOEA Leuckart, 1879. Conoid, if present, forming complete cone; reproduction generally both sexual and asexual; typically with oocysts containing infective sporozoites resulting from sporogony; locomotion of mature organisms by body flexion, gliding, or undulation of longitudinal ridges; flagella present only on microgametes of some groups; pseudopodia ordinarily absent, if present used for feeding, not locomotion.

Subclass GREGARINIA Dufour, 1828. Mature gamonts large, extracellular; mucron or epimerite in mature organism; mucron formed from conoid; generally syzygy of gamonts; gametes usually similar (isogamous) or nearly so, with similar numbers of male and female gametes produced by gamonts; zygotes forming oocysts within gametocysts; life cycle characteristically consisting of gametogony and sporogony; in digestive tract or body cavity of invertebrates or lower chordates; generally one host. *Gregarina, Monocystis.*

Subclass COCCIDIA Leuckart, 1879. Life cycle characteristically consisting of merogony (=schizogony), gametogony, and sporogony; most species in vertebrates, some or all stages intracellular; some species have invertebrate hosts also.

Order EUCOCCIDIIDA Léger and Duboscq, 1910. Merogony present; in vertebrates and/or invertebrates.

Suborder ADELEINA Léger, 1911. Macrogamete and microgamont usually associated in syzygy during development; microgamont producing 1–4 microgametes; sporozoites enclosed in envelope; no endodyogeny; homoxenous or heteroxenous. *Adelea, Haemogregarina, Klossiella, Hepatozoon.*

Suborder EIMERIINA Léger, 1911. Macrogamete and microgamont developing independently; microgamont typically producing many microgametes; zygote not motile; sporozoites typically enclosed in sporocyst within oocyst; one or two vertebrate hosts. *Eimeria, Isospora, Sarcocystis, Besnoitia, Hammondia, Frenkelia, Toxoplasma, Cryptosporidium, Lankesterella, Schellackia.*

Suborder HAEMOSPORINA Danilewsky, 1885. Macrogamont and microgamont developing independently in erythrocytes of vertebrates; conoid ordinarily absent; microgamont producing eight flagellated microgametes; zygote motile (ookinete); sporozoites naked, with three-membraned wall; merogony in erythrocytes of vertebrates and sporogony in blood-sucking insect vectors. *Akiba, Anthemosoma, Babesiosoma, Dactylosoma, Haemocystidium, Haemoproteus, Hepatocystis, Leucocytozoon, Nycteria, Parahaemoproteus, Plasmodium, Polychromophilus, Saurocytozoon, Simondia.*

Subclass PIROPLASMIA Levine, 1961. Piriform, round, rod shaped or ameboid; conoid absent; no oocyst, spore, flagellum or (usually) subpellicular microtubules; typically with polar ring and rhoptries, absent from some members, asexual and sexual reproduction, sporogony in invertebrates and sporozoites with single-membraned wall; all known vectors are ticks.

Order PIROPLASMIDA Wenyon, 1926. *Babesia, Theileria, Cytauxzoon.*

Phylum MICROSPORA Sprague, 1977. Unicellular spores, each with imperforate wall, containing one uninucleate or binucleate sporoplasm and simple or complex extrusion apparatus always with polar tube and polar cap; obligatory intracellular symbionts in nearly all major animal groups.

Class MICROSPOREA Delphy, 1963.

Order MICROSPORIDA Balbiani, 1882. *Encephalitozoon, Minchinia, Nosema, Thelohania.*

Phylum ASCETOSPORA Sprague, 1978. Spore multicellular (or unicellular?); with one or more sporoplasms; without polar capsules or polar filaments; all symbiotic. *Haplosporidium.*

Phylum MYXOZOA Grassé, 1970, emend. Spores of multicellular origin, with one or more polar capsules and sporoplasms, and one, two, three or (rarely) more valves; all symbiotic.

Class MYXOSPOREA Bütschli, 1881. In body cavity or tissues of cold-blooded vertebrates.

Order BIVALVULIDA Shulman, 1959. Spore wall with two valves. *Ceratomyxa, Henneguya, Kudoa, Myxobolus.*

Phylum CILIOPHORA Doflein, 1901. Simple cilia or compound ciliary organelles typical in at least one stage of life cycle; subpellicular infraciliature present even when cilia absent; two types of nuclei with rare exceptions; binary fission transverse (homothetogenic). *Balantidium, Ichthyophthirius, Nyctotherus, Tetrahymena,* "rumen ciliates".

EVOLUTION

Discussion of the evolution of the protozoa is inevitably hypothetical, since many of them (and certainly all of the parasitic forms) have no skeleton or other hard parts and thus only rarely provide a fossil record of their evolution. Hypotheses of the course of evolution which they have undergone, therefore, have to be based largely on deduction from the similarities and apparent relationships of the surviving groups, with the "missing links" being filled in by judicious guesswork.

It is generally accepted, because of the obvious similarities between the simpler members of both groups, that both animal and plant kingdoms had their origins in a single group of organisms; and it is widely, if not generally, believed that the present-day organisms which have deviated least from this primordial group are the Sarcomastigophora. This belief is based on the facts that

(a) flagella (or cilia) are almost universally present among plants and animals, and
(b) among the Mastigophora, species with chloroplasts and species without them are inextricably mixed.

From this primordial assemblage, evolution presumably proceeded along three main lines: the plants, organisms which retained chlorophyll and thus synthesized their own food; the animals, which adopted the habit of eating the fruits of others' labors or the others themselves; and the fungi, with their saprophytic nutrition. Thus in a sense all animals are parasitic on plants but not, of course, within the commonly used definition of this term.

Within the animal group, evolution continued producing many flagellate and ameboid organisms. The present-day survivors of this process are the Zoomastigophorea, Opalinata, Ciliophora, and the Sarcodina. From these, others branched off. The Apicomplexa probably arose quite early in this process, since they possess both flagellate and ameboid characteristics as well as specializations of their own. The origins of the Microspora and Myxozoa are obscure: although at one time placed near the Apicomplexa they are in fact not particularly closely related to this group and are more likely to have arisen separately from the early sarcodine stock. This possibly occurred fairly close to the offshoot leading to the metazoan Coelenterata.

Ideas about evolution within the various protozoan groups are very speculative, and will be considered only broadly here. Within the Ciliophora and Sarcodina, the symbiotic habit was probably adopted independently by organisms of several different groups. Almost all symbiotic members of these groups are mainly or exclusively intestinal parasites, and it is easy to imagine how they could have entered the host's alimentary canal in food or water and multiplied there; eventualy, mutation and selection would have given rise to organisms better suited to this relatively sheltered existence and, finally, to organisms unable to survive outside it unless protected by a cyst. Subsequent evolutionary "experimentation" by the Sarcodina would have led to forms able to invade tissues, as shown today by the facultatively parasitic genera *Naegleria* and *Acanthamoeba*, and by what are probably the most highly evolved parasitic members of the Sarcodina, *Entamoeba histolytica* and its close relatives. Certain species of the ciliate *Tetrahymena* are probably today in the early stages of symbiosis since they appear to be able to live equally well inside or outside their hosts. Some ciliates have become highly adapted physiologically to a symbiotic life (e.g. those inhabiting the alimentary tracts of ruminants).

Similarly, among the Zoomastigophorea, symbiosis has probably arisen independently in several groups: examples are seen among the flagellates living in the intestines of vertebrates and invertebrates, some of which have by now become considerably specialized (e.g. the hypermastigid flagellates of termites and other insects).

One group of flagellates has specialized *par excellence* in symbiosis – the members of the suborder Trypanosomatina of the order Kinetoplastida. These organisms probably evolved from free-living Zoomastigophorea which became adapted to life within the alimentary canal of primitive invertebrate animals in the Pre-Cambrian period over 500 million years ago. From that time they evolved with their hosts, some becoming symbiotic in nematodes, some in molluscs, others in annelids or insects. The trypanosomatids underwent an evolutionary "explosion" in the insects (for reasons which are impossible even to guess at), diverging at first into two main groups: the promastigote group, which retained the primitive anterior position of the flagellar basal body, and the epimastigote stock, which may well have arisen as an adaptation to life in the viscous contents of an insect's gut. In this latter group the posterior displacement of the flagellar insertion made possible the development of an undulating membrane as an aid to locomotion. The insect trypanosomatids are transmitted by the ingestion of resistant forms passed out in the host's feces.

Some authorities have maintained that the genera *Trypanosoma* and *Leishmania* evolved from flagellates living in the intestines of vertebrates, but it seems more likely that they evolved from forms in the guts of invertebrates. When insects adopted the habit of sucking the blood of vertebrates, which they did at least as early as 40 million years ago in the Oligocene period, their trypanosomatid intestinal symbionts were given the possibility of entry into the vertebrate by contamination of the wound caused by the insect with the latter's feces which contained resistant forms of the trypanosomatid. Some trypanosomatids were able to take advantage of this opportunity, and thus the genera *Leishmania* and *Trypanosoma* evolved – the former probably from the promastigote stock, and at least most species of the latter from the epimastigote stock. Some parasitologists believe that *T. cruzi* may have developed from promastigote stock; however, the forms of *T. cruzi* growing in the insect gut are epimastigotes, which suggests that *T. cruzi* arose from epimastigote stock. The forms of *Leishmania* in insects are, on the other hand, promastigotes.

The species of trypanosomes which inhabit aquatic vertebrates and leeches presumably evolved directly from the trypanosomatids of annelids. When certain insects adopted the habit of feeding on plant juices, some of their promastigote symbionts became adapted to life in plants, thus giving rise to the genus *Phytomonas*. This development involved the adoption by the flagellates of transmission via the insect's proboscis, a route also adopted, presumably independently, by members of the genus *Leishmania* and by some of the species of trypanosomes which infect mammals. Alternatively, these (salivarian) trypanosomes may have evolved from leech transmitted trypanosomes of reptiles.

Among the Apicomplexa, the other large group of protozoa which has specialized in symbiosis and of which some members have adopted the hematozoan or blood-dwelling habit, evolution has probably proceeded broadly as follows.

The subclass Gregarinia, all of which live in invertebrates, presumably represents the basal stock of the Apicomplexa. From this stock evolution probably proceeded in a single main line to give rise to those organisms of which the subclass Coccidia are the survivors. It was in this subclass that life within vertebrates was first embarked upon by Apicomplexa. The Coccidia are divided into three groups (ignoring the Protococcida) – Adeleina, Eimeriina and Haemosporina. Presumably the earliest representatives of the two former groups were, as many still are, monoxenous symbionts (i.e. with only one host, invertebrate or vertebrate, playing a part in their life cycle). At first the protozoa inhabited the cells of the host's alimentary canal; but later some of those that were living in vertebrates penetrated deeper into the body of the host, and some became adapted to spending part of their life cycle in the circulating blood cells. This development was associated with the adoption of a heteroxenous life cycle. This type of cycle, which added a phase in a blood-sucking invertebrate vector to the original phase within the vertebrate, increased the organism's chances of finding a new host and was thus of considerable survival value. This sequence of events appears to have occurred at least twice within the Apicomplexa. It occurred among the Adeleina, in those heteroxenous genera in which the sexual individuals (gamonts or gametocytes) and sometimes asexual stages also are found in blood cells, and also among the eimeriines. In the latter group, most genera are still monoxenous, but in those few that have heteroxenous life cycles, such as *Lankesterella*, the transmissive organisms (sporozoites) are found in blood cells; all other stages occur within various fixed tissue cells of the vertebrate host.

The heteroxenous adeleine type of development, in which sexual forms live in the blood cells and fertilization and post-zygotic multiplication (sporogony) occur in an invertebrate vector, is probably the type from which the Haemosporina (malaria parasites, etc.), all of which are heteroxenous, evolved. In the heteroxenous Adeleina and Haemosporina, the part of the life cycle which is undergone in the invertebrate host (sporogony) is that part which, in the monoxenous, more "primitive" Coccidia such as *Adelea* and *Eimeria*, occurs outside the body of the host or in the lumen of its gut. This supports the view of many malariologists, that the monoxenous Apicomplexa which were directly ancestral to the Haemosporina were symbionts of vertebrates – in contrast to the situation postulated above for the evolution of the Trypanosomatina. However, other authorities incline to the opposite view – that the Haemosporina evolved directly from monoxenous Coccidia with invertebrate hosts.

Among the Haemosporina, it is generally thought that *Plasmodium* is the most recently evolved genus. This conclusion is based on the assumption that merogony in erythrocytes is a relatively recent acquisition. The malaria parasites of birds and reptiles probably diverged fairly early from those of mammals. On the basis of the evolutionary sequence of their hosts, the species of *Plasmodium* infecting primates are thought to be the most recently evolved members of the genus; and among this group, probably the "quartan" species (those with a 72 hour cycle of erythrocytic merogony) are the representatives of the first to evolve, while, possibly, *P. falciparum* is the latest. However, the unity of the species infecting mammals is cast in doubt on the basis of DNA analysis, which indicates that *P. falciparum* may be more similar to rodent and avian malarias than to the other primate malarias.

The views expressed above are summarized in the phylogenetic "tree" shown in Figure 1.1.

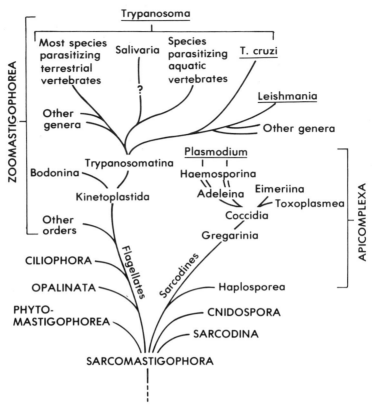

Figure 1.1 Simplified phylogenetic "tree" of the protozoa, representing their presumed evolutionary development.

FURTHER READING

Baker, J. R. 1977. Systematics of parasitic protozoa. In *Parasitic protozoa*, Vol. 1, J. P. Kreier (ed.), 35–56. New York: Academic Press.

Baker, J. R. 1982. Evolution and taxonomic relationships: protozoa. In *Parasites, their world and ours*, D. F. Mettrick & S. S. Desser (eds), 159–68. New York: Elsevier Biomedical Press.

Canning, E. U. 1982. The evolutionary and taxonomic relationships of microspora and myxozoa. In *Parasites, their world and ours*, D. F. Mettrick & S. S. Desser (eds), 175–8. New York: Elsevier Biomedical Press.

Corliss, J. O. 1984. The kingdom Protista and its 45 phyla. *Biosystems* **17**, 87–126.

Cox, F. E. G. 1981. A new classification of the parasitic protozoa. *Protozoological Abstracts* **5**, 9–14.

Honigberg, B. M. 1982. Evolutionary and taxonomic relationships among

Zoomastigophorea. In *Parasites, their world and ours*, D. F. Mettrick & S. S. Desser (eds), 172–4. New York: Elsevier Biomedical Press.

Landau, I. 1982. Hypothèses sur la phylogénie des coccidiomorphes de vertebrés; évolution des cycles et spectre d'hôtes. In *Parasites, their world and ours*, D. F. Mettrick & S. S. Desser (eds), 169–71. New York: Elsevier Biomedical Press.

Levine, N. O., J. O. Corliss, F. E. G. Cox, G. Deroux, J. Grain, B. M. Honigberg, G. F. Leedale, A. R. Loeblich, J. Lom, D. Lynn, E. G. Merinfeld, F. C. Page, G. Poljansky, V. Sprague, J. Vavra & F. G. Wallace 1980. A newly revised classification of the protozoa. *Journal of Protozoology* **27**, 37–58.

McCulchan, T. F., J. B. Dame, L. H. Miller & J. Barnwell 1984. Evolutionary relatedness of *Plasmodium* species as determined by the structures of DNA. *Science* **225**, 808–11.

Whittaker, R. H. 1977. Broad classification: the kingdoms and the protozoans. In *Parasitic protozoa*, Vol. 1, J. P. Kreier (ed.), 1–34. New York: Academic Press.

Woo, P. T. K. 1970. Origin of mammalian trypanosomes which develop in the anterior-station of blood sucking arthropods. *Nature* **228**, 1059–62.

CHAPTER TWO

Anatomy and physiology of the protozoa

ANATOMY

The protozoa are structurally equivalent to a single metazoan cell: basically, a mass of cytoplasm bounded by some kind of limiting membrane and containing one or more nuclei. The cytoplasm may contain most, if not all, of the organelles found in metazoan cells, including mitochondria, endoplasmic reticulum, ribosomes, Golgi apparatus, bodies resembling lysosomes, fibrils and microtubules of various kinds, centrioles, flagella and cilia (Fig. 2.1). Not all protozoa, of course, possess all these organelles. Other structures, apparently exclusive to protozoa, are sometimes present. These are of fewer kinds; examples include the trichocysts of *Paramecium* (a free-living ciliate), various skeletal structures, and, perhaps, the contractile vacuoles. Some of these organelles will now be considered in a little more detail.

Nucleus

There is no basic difference between the nucleus of a protozoon and that of a metazoon. Both are usually spherical or disc-shaped organelles, bounded by a double unit membrane which is, in some places, continuous with the endoplasmic reticulum (Fig. 2.2). Protozoan nuclei contain deoxyribonucleoprotein and ribonucleoprotein; the latter is often concentrated into one or more intranuclear masses called nucleoli. Chromosomes have been reported from the nuclei of some protozoa (Fig. 2.3), but the small size of the nuclei makes their demonstration difficult.

The chromosome numbers recorded for most parasitic protozoa are low, and most reports are probably suspect because of technical limitations. However, recent (1985) studies using pulsed-field gradient gel electrophoresis identified seven (haploid) chromosomes in *Plasmo-*

EUKARYOTIC CELL

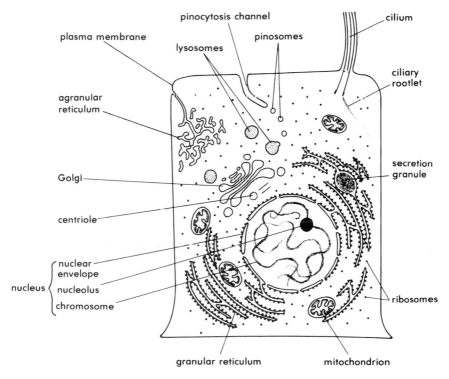

Figure 2.1 Diagram showing structure of generalized eukaryotic cell to demonstrate nuclear and cytoplasmic organization (from Vickerman & Cox 1967).

dium falciparum and 20 or more, plus about 100 'minichromosomes', in *Trypanosoma brucei* (which is probably diploid). The deoxyribonucleic acid (DNA) content of the nuclei of parasitic protozoa is also low in the few organisms for which it has been determined: in *Trypanosoma brucei* the nuclear DNA content is only about three percent of that found in human diploid nuclei, and in *T. brucei* the total DNA content of each organism has been estimated at about 0.1 ng.

Ciliophora have two nuclei: one sexual nucleus (the micronucleus) and one polyploid asexual nucleus (the macronucleus) (Fig. 2.4). A few protozoa have many nuclei (e.g. the Opalinata), but the majority are uninucleate. Among symbiotic protozoa, sexual processes are known only among the Ciliophora, Mastigophora and Apicomplexa: in the first, the micronucleus is diploid throughout most of the organism's life history, while in some (perhaps all) Apicomplexa the nucleus is haploid for all but a brief period of time after fertilization.

13

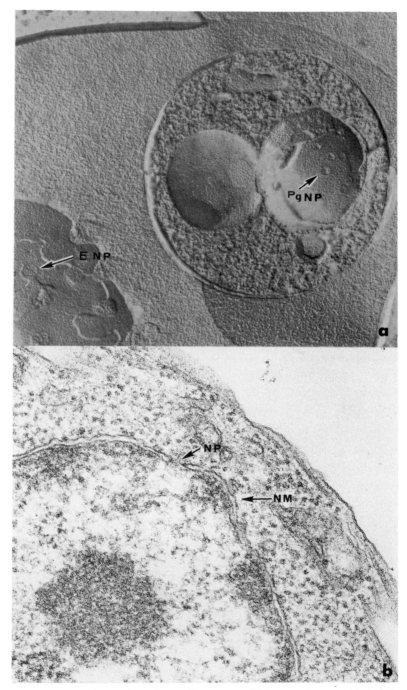

Figure 2.2 (a) Binucleate *Plasmodium gallinaceum* in chicken erythrocyte. Freeze-cleaved preparation showing nuclear pores (PgNP). Nuclear pores are also visible in the chicken erythrocyte nucleus in the lower left corner of the photograph (ENP). (b) Thin section electron micrograph of a *Trypanosoma cruzi* epimastigote showing double nuclear membrane (NM) and nuclear pore (NP).

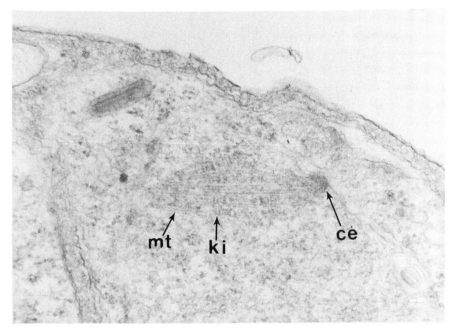

Figure 2.3 Electron micrograph of a mitotic figure in *Plasmodium berghei* showing intranuclear microtubules (mt). The microtubules extend from centrioles (ce). Small electron-dense bars, possibly kinetochores (ki), are present attached to the microtubules (micrograph provided by Dr Xu Linhe, Department of Parasitology, Shanghai Medical University, People's Republic of China).

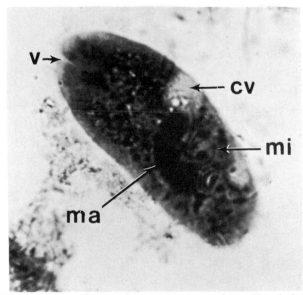

Figure 2.4 Ciliate showing macronucleus and micronucleus. The dark bean-shaped structure is the macronucleus (ma); the small round structure within the inner curve is the micronucleus (mi). A contractile vacuole (cv) and vestibule (v), where ingestion occurs, are also visible. The organism is *Balantidium coli*, a parasite in the colon of pigs and, rarely, man.

Mitochrondria

Basically these are similar in structure to those of most metazoa, though often the infoldings of the inner membrane (cristae) are tubular or discoidal rather than the flat ridge-like structures seen in metazoa (Fig. 2.8). However, some piroplasmids have non-cristate mitochondria (Fig. 2.5). Many parasitic protozoa have mitochondria at least during some stage of their life cycle, with the exception of anaerobic forms inhabiting the host's alimentary canal, from which mitochondria may have been secondarily lost. Some species of malaria parasites (e.g. *Plasmodium berghei*) do not have conventional mito-chondria in their asexual stages in the host's erythrocytes, but possess, instead, organelles composed of concentric membranes which may serve the same purpose. However, in other stages of their life cycle, typical protozoan mitochondria are present (Fig. 2.6). Certain Mastigo-phora (the order Kinetoplastida, most of which are symbiotic) have an unusually large amount of mitochondrial DNA aggregated as one or more structures called kinetoplasts (Fig. 2.7) inside their single, large mitochondrion.

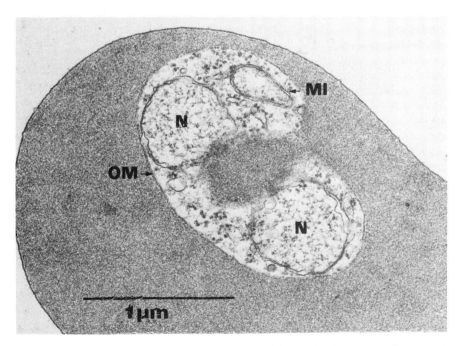

Figure 2.5 Thin section electron micrograph of *Babesia microti* in an erythrocyte. A noncristate mitochondrion (MI) and two double membrane-bound nuclei (N) are visible (from Kreier *et al.* 1975).

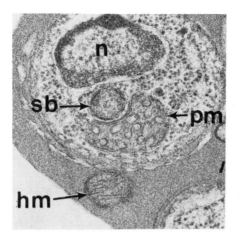

Figure 2.6 This micrograph of a *Plasmodium* species inside an erythrocyte shows the nucleus (n), a parasite mitochondrion (pm), a host cell mitochondrion (hm), and a spherical body (sb). The spherical body is of unknown function (photomicrograph supplied by Dr Masamichi Aikawa, Case Western Reserve University).

Secretory organelles

Under this heading are grouped the rough and smooth endoplasmic reticulum and the ribosomes (Fig. 2.8), which are present in many if not all protozoa. The smooth endoplasmic reticulum is often specialized, as in metazoan cells, into a pile of flattened sacs variously known as a Golgi apparatus or dictyosome. In some Mastigophora the structure termed "parabasal body" is equivalent to a Golgi apparatus. In certain older textbooks, the term "parabasal body" was used for the kinetoplast, a totally distinct structure (see above).

Fibrils and microtubules

Simple microtubules (composed, as far as is known, of the protein tubulin) have been seen in many protozoa, often underlying the limiting membrane of those organisms which have a fairly definite shape – perhaps they serve to maintain this shape (Fig. 2.9). Intranuclear spindle microtubules have also been seen in some species, and may well be of general occurrence. The development of microtubules and fibrils seems, among the protozoa, to have reached its maximum in the Ciliophora.

17

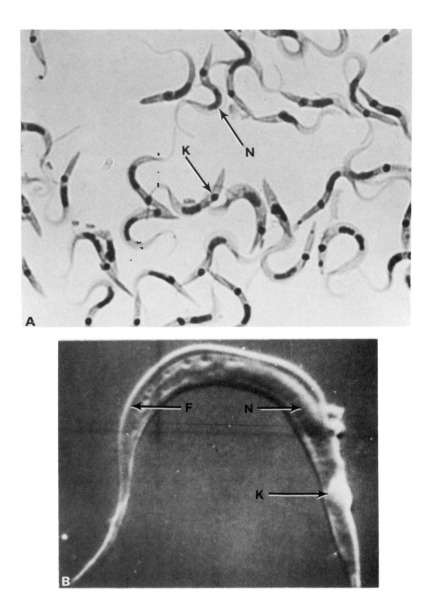

Figure 2.7 (a) Photomicrograph of *Trypanosoma cruzi* trypomastigotes stained with Giemsa's stain. The deeply staining spherical kinetoplasts (K) and the elongate nuclei (N) are clearly visible in the less intensely stained organisms. (b) Scanning electron micrograph of a *T. cruzi* trypomastigote. The spherical kinetoplast (K), flagellum (F), and elongate nucleus (N) are visible.

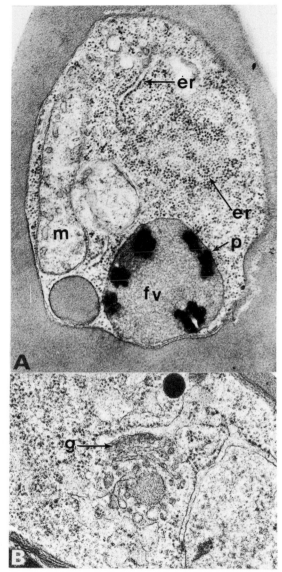

Figure 2.8 These micrographs are of *Plasmodium*. In (a) polysomes, which make up, in part, the endoplasmic reticulum (er), are seen as electron-dense dots. A food vacuole (fv) containing malarial pigment (p) and a mitochondrion (m) are also present. In (b) the stacked membranous structures may form a Golgi apparatus (g) (photomicrographs supplied by Dr Masamichi Aikawa, Case Western Reserve University).

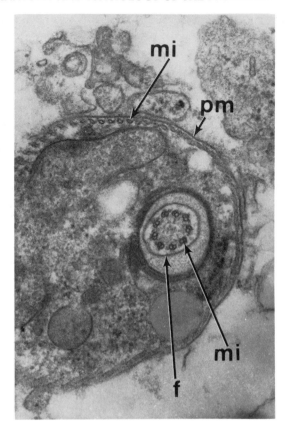

Figure 2.9 Electron micrograph of a *Leishmania mexicana* amastigote. Microtubules (mi) are present just under the plasma membrane (pm) and also in the cross section of the flagellum (f) (micrograph provided by Dr Xu Linhe, Department of Parasitology, Shanghai Medical University, Shanghai, People's Republic of China).

Flagella and cilia

These, as their names imply, are whip-like or hair-like filamentous extensions from the body surface of many protozoa (and metazoan cells). Both are fundamentally identical structures and perhaps should not really be given different names. They are quite different from prokaryotic flagella, and it has been suggested that this distinction should be emphasized by introducing a new term (e.g. "undulipodium") for the eukaryote organelle. They are contractile, and beat or wave to and fro in a variety of complex patterns. The organelles conventionally termed cilia are usually shorter and more numerous than those called flagella, and normally beat in a synchronous, progressive pattern called metachronal rhythm, aptly described as resembling the appearance of a field of wheat waving in the breeze.

All eukaryotic cilia and flagella (of protozoa, metazoa, and even plants) have an identical basic structure – a cylinder of nine double

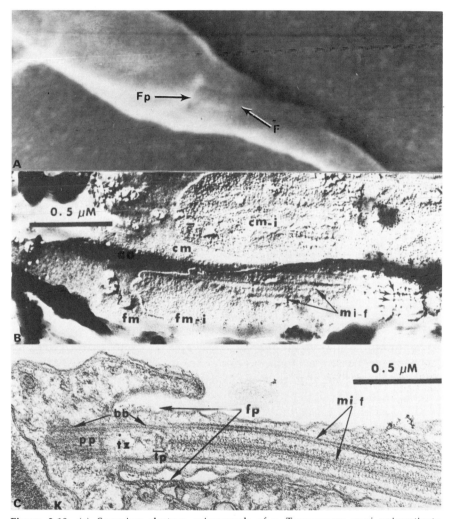

Figure 2.10 (a) Scanning electron micrograph of a *Trypanosoma cruzi* epimastigote showing the flagellum (F) entering the flagellar pocket (Fp). (b) Electron micrograph of the flagellum and body of a freeze-cleaved *T. cruzi* epimastigote. Interior and exterior surfaces of the cytoplasmic and flagellar membrane (cm-i; cm; fm-i; fm) are visible, as are the microtubules of the flagellum (mi-f). (c) Electron micrograph of a longitudinal section through the region at the base of the flagellum of *T. cruzi*. The kinetoplast (K), flagellar microtubules (mi-f) and flagellar pocket (fp) can be seen, as can the basal body (bb) with its component proximal portion (pp), transition zone (tz) and terminal plate (tp) (from Seed *et al.* 1973).

21

microtubules (each resembling a figure 8 in cross section), composed of the protein tubulin, with a central core of two single fibrils, surrounded by a sheath, to which each doublet is connected by a radial "spoke". Each doublet of external fibrils has a pair of lateral movable arms, composed of a protein called dynein, which can link it to its neighbor. This "9+2'" cylinder (the axoneme) is surrounded by a hollow cylindrical extension of the limiting membrane of the organism (or cell), the sheath. The flagella of some protists have a band of proteinaceous material within the sheath, parallel to the axoneme, which perhaps has a supporting function. Both flagella and cilia arise from a basal body or kinetosome (sometimes called "blepharoplast") within the cell. The basal body consists essentially of a relatively short extension of the nine outer axoneme fibrils (which may here be triple instead of double), without the two central ones which often appear to originate at a basal plate marking the junction of axoneme and basal body (Fig. 2.10). The basal body appears to function as a template for assembly of the axonemal microtubules. It is homologous with the centriole, and indeed sometimes also functions as a centriole during nuclear division (in, for example, the green protist *Chlorogonium*).

In many Mastigophora the flagellar sheath is attached to the organism's limiting membrane by desmosome-like structures and, as the flagellum beats, the membrane is pulled out to form a fin-like undulating membrane (Fig. 2.11). Since both cilia and flagella are primarily locomotory organelles in protozoa, the undulating membrane is presumed to help in this function. Certain cilia and flagella have a secondary function as food-gathering organelles – they draw water currents containing food particles into special gullet-like grooves or invaginations of the protozoon's body surface. The propulsive effect of cilia has been likened to that of the oars used in rowing a boat. Some flagella function similarly, while others act as a propeller either pushing or pulling (depending on the direction in which the wave of contraction passes along the flagellum) the organism through its liquid environment. In protozoa with only one or few flagella, the latter usually arise anteriorly, though one or more may be recurved to run back along the organism's surface (often forming an undulating membrane, as described above); in some Trypanosomatina, the flagellar origin has moved posteriorly.

Limiting membrane

Presumably the basic limiting membrane of cells is the unit membrane, and this is found as the plasmalemma of many protozoa (e.g. the trypanosomes and amebae). Outside the plasmalemma an additional layer, such as a glycoprotein "surface coat" or glycocalyx, may be

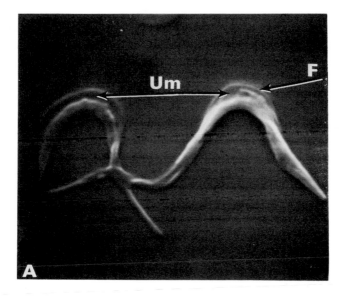

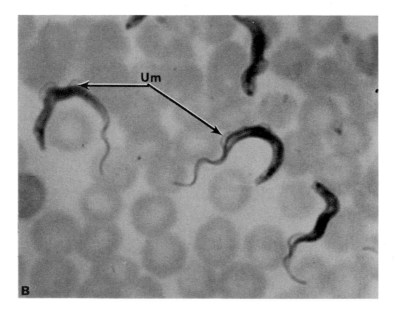

Figure 2.11 The undulating membrane (Um) is formed by attachment of the flagellum (F) to the body of the organism. In (a) the flagellum (F) can be seen adherent to the body of the trypanosome (scanning electron micrograph of *Trypanosoma cruzi*). In (b) the undulating membrane (Um) appears as loops along the body of the trypanosome (Giemsa-stained thin blood film containing *Trypanosoma brucei*).

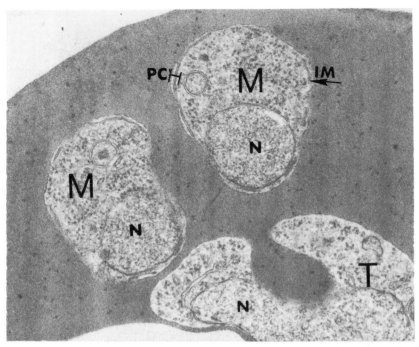

Figure 2.12 Thin section electron micrograph of *Babesia microti* merozoites (M) and trophozoite (T) in an erythrocyte. Parasite nuclei (N) are visible, and a pellicle complex (PC), in which the plasmalemma is reinforced by an inner membrane (IM), is present in the merozoites but not the trophozoites (from Kreier *et al.* 1975).

present. However, in other groups, e.g. motile stages of malaria parasites and piroplasms, a more complicated membrane has developed, consisting of two or more unit membranes (Fig. 2.12), sometimes apparently with additional material between them – perhaps to give added rigidity. In the Ciliophora a complex pellicle is present, often consisting of an outer unit membrane with a layer of sacs or alveoi beneath it. In some species (e.g. *Paramecium*) the sacs are inflated, producing a "sculptured" effect on the pellicular surface, which may be presumed to confer rigidity.

The various kinds of protective walls secreted by many protozoa may be considered as derivatives of the limiting membrane. Among the symbiotic protozoa, the resistant forms called cysts by protozoologists are commonly produced by those genera which inhabit their host's alimentary canal and spend part of their life cycle outside the host while awaiting ingestion by a new host (e.g. symbiotic amebae, and the monoxenous coccidia belonging to the genera *Eimeria*, *Isospora*, etc.). These cysts serve to protect the organism from desiccation or other damage during this vulnerable part of its life cycle (Fig. 2.13).

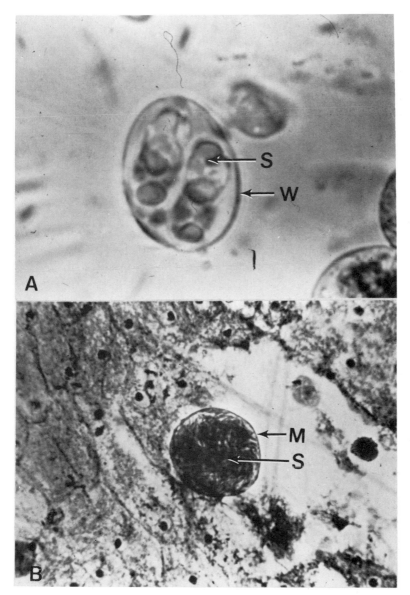

Figure 2.13 (a) A sporulated coccidial oocyst with a thick cyst or spore wall (W) which serves to protect the enclosed infective sporozoites (S) when the oocyst is passed out with the feces. (b) The oocyst of *Plasmodium* is surrounded by a tough membrane (M) which protects the developing sporozoites (S) from contact with the mosquito's defenses.

Cysts are also produced by some symbiotic protozoa which have developed more efficient means of transmission by vector animals, avoiding the hazards of a "free" existence, such as the oocysts of *Plasmodium*. These cysts may be regarded as an evolutionary legacy from their ancestors. The function of such cysts is probably still protective – but against the host's defenses rather than against a hostile environment outside the host.

The choice of the word "cyst" for the resting, walled form of protozoa was in some respects unfortunate. The term is used by pathologists to mean a fluid-filled sac which may be produced by the body as a result of some irritation. Resting walled forms are called spores by bacteriologists and mycologists; that term would be appropriate for the equivalent forms of protozoa, and is indeed used for some such forms (e.g. of Microspora and Myxozoa). However, to confuse the issue still further, the term "cyst" is used to refer to two distinct structures in the life cycle of *Toxoplasma*: the cyst (oocyst) produced following gametic fusion in the definitive host, and the structure containing the zooites in the tissues of other hosts.

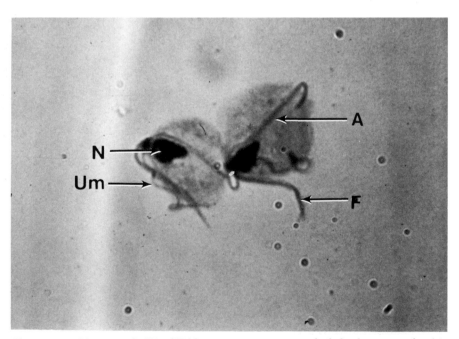

Figure 2.14 The axostyle (A) of *Trichomonas* serves as an endoskeletal structure for this organism. The nucleus (N), flagellum (F) and an undulating membrane (Um) are also visible in this photomicrograph of two fixed, stained *T. vaginalis*.

Skeletal structures

Many free-living protozoa have elaborate exoskeletons (e.g. Radiolaria, Foraminifera), but these are absent from symbiotic forms (unless cyst walls are considered as exoskeletal structures). Perhaps the tough walls which surround the spores of some Myxozoa and Microspora may legitimately be regarded as exoskeletal. Endoskeletal structures are found in some symbiotic protozoa, particularly the Mastigophora (e.g. the axostyle of *Trichomonas*), but these, too, are relatively rare (Fig. 2.14). As mentioned above, the subpellicular microtubules of trypanosomatids and malaria parasites may be endoskeletal in function.

Contractile vacuoles

As the primary function of these organelles is to remove unwanted water entering the organism by osmosis or during feeding, they are rarely found in symbiotic protozoa, which usually inhabit an isotonic environment. However, they are seen in symbiotic ciliates, and in the amebae *Acanthamoeba* and *Naegleria* which sometimes live in man (Fig. 2.15). A contractile vacuole consists of a rhythmically pulsating vesicle, fed by a system of radial canals and opening to the outside through a small pore in the organism's limiting membrane. A complicated

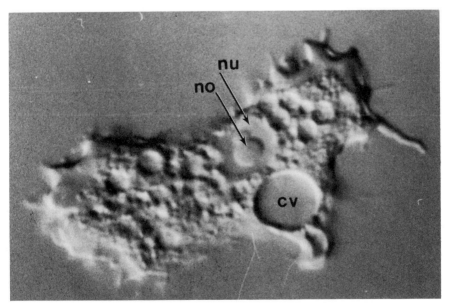

Figure 2.15 *Acanthamoeba castellanii*. This free-living ameba has at least one contractile vacuole (cv) which aids it in osmoregulation. A prominent nucleolus (no) is present in the nucleus (nu) (micrograph provided by Dr T. Byers, Ohio State University).

arrangement of "valves" ensures that liquid flows in only one direction – from the canals into the vacuole and then, when the latter contracts, out through the pore.

Apical complex

At the anterior end of the organism in at least some stages in the life cycle of most Apicomplexa there is a characteristic assemblage of

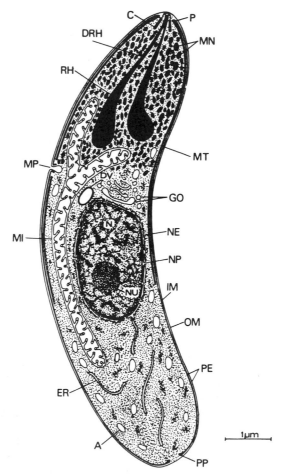

Figure 2.16 Diagram of the structure of a coccidian merozoite. The apical structures are characteristic of the motile stages of the Apicomplexa. A, amylopectin; C, conoid; DRH, ducts of rhoptries; ER, endoplasmic reticulum; GO, Golgi apparatus; IM, inner membrane; MI, mitochondrion; MN, micronemes; MP, micropore; MT, microtubules; NE, nuclear envelope; NP, nuclear pore; OM, outer membrane; P, polar ring; PE, pellicle; PP, posterior polar ring; RH, rhoptries (from Scholtyseck 1979).

structures (Fig. 2.16) from which this phylum received its name. These are discernible only by electron microscopy (Fig. 2.17), and not all are necessarily present in all Apicomplexa. The full complement, shown diagramatically in Figure 2.16, consists of the following parts:

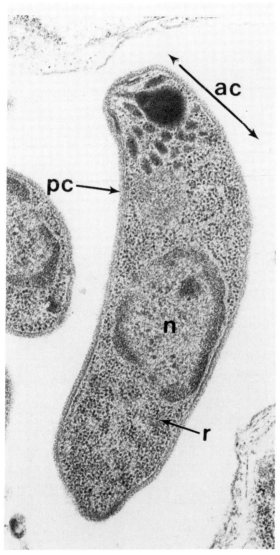

Figure 2.17 This electron micrograph shows an exoerythrocytic merozoite of *Plasmodium*. The apical complex (ac), pellicle complex (pc), nucleus (n), and ribosomes (r) are clearly visible (photomicrograph supplied by Dr Masamichi Aikawa, Case Western Reserve University).

(a) Polar ring, encircling the anterior tip and acting as microtubule organizing centre (MTOC) for the subpellicular microtubules.

(b) Conoid, an apparently protrusible "nose-cone", possibly constructed of parallel longitudinal components and probably used in penetration of the host cell.

(c) Rhoptries, sac-like bodies with a narrow stem leading to the anterior tip, which are thought to produce a secretion aiding host cell penetration.

(d) Micronemes, convoluted rods or tubules, or possibly multiple shorter structures, extending posteriad for more or less of the length of the cell; their function is unknown, but they are speculatively ascribed a similar role to the rhoptries and conceivably are developmental stages or subsidiaries ("feeders") of the latter.

PHYSIOLOGY

Locomotion

Basically there are three known methods by which protozoa move: (a) ameboid movement, (b) flagellar and ciliary movement, and (c) gregarine movement. These will be discussed briefly one by one.

Ameboid movement   Not surprisingly, this is the method by which amebae, as well as some other protozoans, move. It is characteristic of the whole superclass Sarcodina and is also used by some Mastigophora and few Sporozoea (at certain stages of the life cycle), but not by any of the Ciliophora (presumably the complexity of their pellicle precludes it). Ameboid movement consists of the temporary extension of the body into one or more processes, called pseudopodia ("false feet"), in the direction in which movement is occurring; the rest of the body is then drawn up into the pseudopodium, and the whole procedure is repeated (Fig. 2.18). Pseudopodia may be of different kinds – lobopodia (broad and blunt in shape), filopodia (long and thin), reticulopodia (thin and branching, forming a network) and axopodia (long and thin, with a central supporting rod). The classification of the Sarcodina is based in part on the type of pseudopodium present. All parasitic amebae have pseudopodia basically of the lobopodial type. Species of *Entamoeba* move by means of protrusion of the ectoplasm (the outer, clear layer of cytoplasm) at the advancing end of the organism; as this protrusion extends, the endoplasm (the inner cytoplasmic layer) beings to flow into it, and gradually the whole organism "catches up," as it were, with the advancing front.

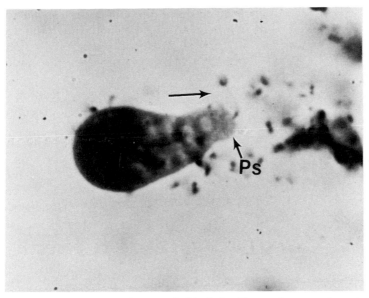

Figure 2.18 Photomicrograph of *Entamoeba histolytica*. The ameba moves by extending a pseudopodium (Ps) and then its body flows into the extension. The direction of movement is indicated by the arrow.

The precise mechanism involved in the production of pseudopodia is still not completely understood, but the most popular theory at present is that it depends upon active contraction of the ectoplasmic "tube" at the hind end of the body: thus the endoplasm is squeezed forwards into the expanding pseudopodium, while the ectoplasm forming the sides of the tube remains stationary. The process must involve continual transformation of ectoplasm to endoplasm and breakdown or absorption of the pellicle at the hind end, and reformation (from endoplasm) at the front. *Entamoeba* species thus have a definite polarity – the hind end, or "uroid", being a relatively constant feature. Whether this polarity remains unchanged throughout the life of the ameba (ie. the period from one cell division to the next) is unknown. There is evidence that adenosine triphosphate (ATP), as an "energy carrier", and proteins resembling actomyosin play a part in ameboid movement, as they do in metazoan muscular contraction.

Flagellar and ciliary movement   The structure of flagella and cilia, and the main types of beat characteristic of each, have been described in the preceding section. Ciliary and flagellar beat is produced by a mechanism involving the sequential sliding of each outer fibril doublet relative to its neighbor, mediated by a reversible coupling of the tips of

31

the dynein arms to the adjacent microtubule of the neighboring doublet. The presence of ATP seems to be essential; dynein is an enzyme which degrades ATP to the diphosphate. The coordination of this sequential rotary sliding is not fully understood, but may depend on interactions between the radial spokes and the central pair of fibrils of the 9+2 cylinder, or their sheath. The mechanism responsible for synchronizing the metachronal rhythm of beating cilia can be fully explained in terms of mechanical effects transmitted from cilium to cilium by pressure waves in the surrounding medium.

Flagellar and ciliary movement is, of course, typical of the Mastigophora and Ciliophora. It is also seen in the Opalinata and a few Sarcodina (at certain stages of the life cycle). Among the Apicomplexa, only microgametes of organisms of the subclass Coccidia and of some members of the subclass Gregarinia are flagellate.

Gregarine movement   This is typical of the Apicomplexa at certain stages of their life cycles. It is probably the least understood of all the types of protozoan locomotion. An organism moving in this way when seen under a light microscope seems merely to glide along without using any special organelles. Gregarine movement may be produced by small-scale contraction or "rippling" of the pellicle, moving the organism in a way similar to that in which a slug moves, though other hypotheses have been suggested.

Movement of this kind (or believed to be of this kind) is shown by many gregarines (hence its name), by gametocytes of eimeriine haemogregarines, by ookinetes of the Haemosporina, by sporozoites of many of the Apicomplexa, and by Toxoplasmea.

Nutrition

There are many schemes for classifying nutritional mechanisms. A simple one, adequate for the present purpose, places organisms into two categories: autotrophs or heterotrophs. **Autotrophs** are those which obtain carbon from carbon dioxide or other one-carbon compounds. The energy required to synthesize their own complex carbon compounds is usually acquired from sunlight; organisms using this process (photosynthesis) are photoautotrophs. No protozoon considered in this book belongs to this category. **Heterotrophs** are organisms which can obtain carbon only from pre-synthesized organic compounds. A few protozoa do this by utilizing light energy (photoheterotrophs) but most (including all symbiotic forms) obtain energy by oxidation; these are chemoheterotrophs. Heterotrophic nutrition consists essentially of the ingestion of complex molecules of proteins, carbohydrates, fats, etc., in the form of other animals or plants (either whole, in pieces, or in solution), breaking these molecules down into simpler units if necessary (amino acids, simple sugars, fatty acids) and rebuilding them into the animal's own proteins, carbohydrates and fats.

There are basically two routes of entry for molecules into the interior of the protozoan cell: passage through the cell membrane, or engulfment into a vacuole consisting of a "pinched-off" portion of that membrane. Small molecules in solution can pass to and fro through the membrane down a concentration gradient by facilitated diffusion, a process which involves carrier proteins in the membrane and does not require energy, or by active transport, an energy-requiring process operating against a concentration gradient and also involving carrier proteins in the membrane. Active transport is an important process in many parasitic protozoa, which live bathed in a liquid medium of high nutrient concentration, such as blood plasma.

Engulfment into a vacuole, or endocytosis, is followed by fusion of the vacuole (endosome or phagosome) with a lysosome and digestion of the vacuole's contents by lysosomal enzymes. The products of this digestion are then presumably absorbed by diffusion or active transport. Depending on its scale, endocytosis may be differentiated into phagocytosis (ingestion of large particles) or pinocytosis (ingestion of smaller volumes of, usually, semi-solid or liquid nutrient materials). Endocytosis may be possible only at a specialized region of the membrane, the cytostome (e.g. in some trypanosomes, many other flagellates, and ciliates); it may be limited to certain specialized areas (e.g. the flagellar pocket of some trypanosomes); or it may occur generally over all the surface (e.g. in some amebae).

Endocytosis may be initiated at membrane receptor sites specific for the object to be engulfed. In cultured mammalian cells, at least, pinocytosis appears to begin at membrane indentations called coated pits, which become coated vesicles in the cytoplasm before delivering their contents to endosomes; it is possible that a similar process may operate in Protista too.

Symbiotic protozoa show a trend towards dietary specialization, presumably with concomitant production of fewer digestive enzymes, compared to their free-living relatives. This is reflected in the fact that they are usually more difficult to cultivate *in vitro* in non-living media than are free-living forms. At one end of the scale are some of the parasitic amebae and flagellates, almost, if not quite, as easy to cultivate as their free-living relatives; at the other end are the obligate intracellular forms, such as *Toxoplasma* and (for most of its life cycle) *Plasmodium*, which are very difficult even to keep alive, let alone to grow, outside a suitable living cell.

The food-gathering mechanisms used by the symbiotic protozoa are generally the same as those of their free-living relatives. Among the Zoomastigophorea, some of the forms inhabiting the host's intestine have a distinct mouth, or cytostome, through which quite large food particles are ingested (phagotrophy). The cytostome is at the base of a groove, or cytopharynx, containing a flagellum, which produces water currents to draw food particles down the groove. A similar mechanism is employed by many symbiotic Ciliophora. The Zoomastigophorea which inhabit the host's blood or tissue fluids (the trypanosomes) feed rather similarly, but do not have a single, relatively large mouth. Instead, small (submicroscopic) droplets of the host's plasma are ingested by pinocytosis or through a cytostome within the flagellar invagination (a deep intucking through which the proximal part of the flagellum passes).

Probably all Sporozoea feed by phagotrophy, at least in part and at certain stages of their life cycle, through one or more cytostomes (small cylindrical depressions) in the pellicle. In Sporozoea the pellicle consists of at least two unit membranes, and only at the base of the cytostome is the cytoplasm bounded by a single unit membrane.

Symbiotic Rhizopodea of the genus *Entamoeba* feed by phagotrophy, taking in food particles at the hind end (uroid), where the pellicle of the organism seems to be sticky. Particles adhere to the hind end and become drawn into the ameba's cytoplasm when the pellicle is drawn in at this region during the process of locomotion. This is rather different from the engulfment of food particles by pseudopodia which is the way in which most free-living Rhizopodea catch their food. It is uncertain which method is adopted by the other genera of symbiotic amebae.

It is not known to what extent diffusion or active transport across the surface membrane play a part in the ingestion of nutrients by symbiotic protozoa, in addition to the processes outlined above.

The digestive processes by which the protozoa break down the nutrients which they ingest are little known, but are probably similar, at least in outline, to the hydrolytic processes used by the metazoa. Many symbiotic protozoa acquire their nutrients already in the form of simple molecules, digestion having been done for them by the host, and further breakdown is not necessary.

Respiration

The purpose of feeding is to obtain raw material for growth and repair, and also to obtain fuel from which the energy required by the cells can be released. The processes by which energy is released are called respiration. Respiration may be aerobic (requiring oxygen) or anaerobic. Examples of both are found among the symbiotic protozoa.

Generally, those living in the alimentary canal of vertebrates, where the oxygen tension is low, are anaerobic.

As far as is known, aerobic respiration in almost all symbiotic protozoa follows the same general pattern as it does in metazoan cells, being based on the oxidation of glucose to carbon dioxide and water via the Embden–Meyerhof pathway, Krebs' tricarboxylic acid cycle and cytochrome systems, the two latter systems being (again as in metazoal cells) associated with the mitochondria. However, the stages of *Trypanosoma brucei* which multiply in the blood of the mammalian host (but not those which develop in the insect vector) have adopted an apparently unique non-mitochondrial system of aerobic respiration. The initial stages of glycolysis (Embden–Meyerhof pathway) are the same, but glucose is degraded only as far as pyruvic acid, which is excreted. The subsequent stages of oxidation, to carbon dioxide and water, and oxidative phosphorylation (involving Krebs' cycle and cytochrome system), both of which normally take place in mitochondria, do not occur. Instead, terminal respiration is mediated by an L-α-glycerophosphate oxidase–L-α-glycerophosphate dehydrogenase system, located in small membrane-bound vesicles (hydrogenosomes) which are found throughout the cytoplasm.

Carbohydrate storage products vary widely, with the polysaccharides starch or glycogen being the commonest. *Entamoeba* stores glycogen, as do some intestinal flagellates; *Eimeria* stockpiles amylopectin. Inhabitants of the carbohydrate-rich environment provided by blood (e.g. *Plasmodium* and *Trypanosoma*) do not need carbohydrate reserves.

Excretion

The excretion of soluble waste products from all protozoa usually occurs by diffusion. Protozoa which have contractile vacuoles doubtless remove some soluble waste via these organelles, but their main function is osmoregulatory.

Insoluble matter is ejected from the food vacuoles (endosomes) of Ciliophora through a small pore (cytopyge) in the pellicle. Rhizopodea behave similarly but, since they are bounded by only a single unit membrane, do not seem to require a special pore.

The intraerythrocytic stages of malaria parasites (species of *Plasmodium*) ingest host cell cytoplasm by endocytosis through a cytostome. The hemoglobin contained in this cytoplasm is only partially digested. The protein component (globin) is degraded to its constituent amino acids, while the iron-containing heme moiety is converted to insoluble brown or black "malaria pigment" (hemozoin), probably a compound of hematin and a polypeptide. The hemozoin then remains parcelled up in the parasite's endosomes, and is left behind at the next division process. This system has the merit of removing free hematin from the protozoan's cytoplasm, hematin being an inhibitor of succinic dehydrogenase, a Krebs' cycle enzyme.

Asexual reproduction

This can occur in one of at least five ways.

Binary fission   This is the simple division of one individual into two and is the commonest form of asexual multiplication. The organism's nucleus first divides by mitosis. The details of nuclear division have been elucidated completely in very few protozoa, partly because of the small size of the nuclei. In some, but not all, protozoa mitosis occurs within the intact nuclear membrane (differing from the process in metazoa): chromosomes, spindle fibrils, etc., have been seen during some protozoan mitoses and are probably generally, if not universally, formed (see Fig. 2.3). There is doubt as to precisely how the polyploid macronuclei of the Ciliophora divide.

Following nuclear division (karyokinesis), the organism's cytoplasm divides (cytokinesis): in most groups (Rhizopodea, Ciliophora, and Sporozoea) this appears to result from the development and deepening of a furrow around the organism in the plane of fission. In Ciliophora and those Sporozoea which undergo binary fission, division is equatorial or transverse (division being said to be homothetogenic in these groups). In the Mastigophora, the plane of fission is meridional or longitudinal (division being said to be symmetrogenic), and the separation of the two "daughter" organisms begins anteriorly and proceeds steadily back (Fig. 2.19). Some organelles divide at the time of division (e.g. nucleus, mitochondria presumably), while others have to be formed anew by one of the "daughters" (e.g. flagella, cytostomes, and the ciliary mouth complexes of Ciliophora).

Multiple fission   This is a variant of binary fission in which a subsequent division commences before the earlier one is completed. It is seen particularly among Eugregarinida (Apicomplexa), resulting in long chains of individuals, and in Trypanosomatina (Zoomastigophorea), where it may result in the formation of spheres or rosettes of organisms all still attached at their hind ends.

Budding   In this type of division, a new, small individual develops as a bud on the surface of the old (see Fig. 8.2a, b), one of the products of nuclear division entering the bud (Fig. 2.20b); finally the bud breaks off. This method is common among certain Ciliophora and piroplasms but rare among other symbiotic protozoa.

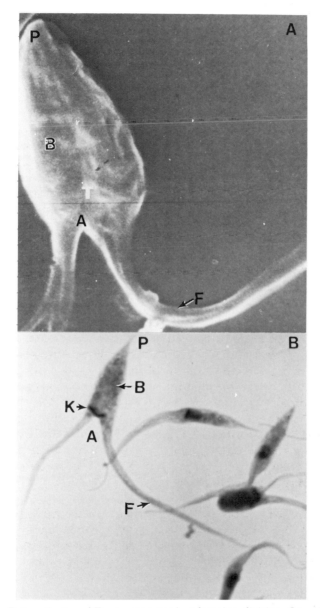

Figure 2.19 Epimastigotes of *Trypanosoma cruzi* undergoing division. Scanning electron micrograph (a) and photomicrograph of Giemsa-stained organisms (b). Division of the body (B) proceeds from the anterior end (A) to the posterior (P). The flagellum (F) has already been duplicated and a division trough (T) is visible in the body of the organism. The kinetoplast (K), in the process of division, is visible in the photomicrograph of the Giemsa-stained organism.

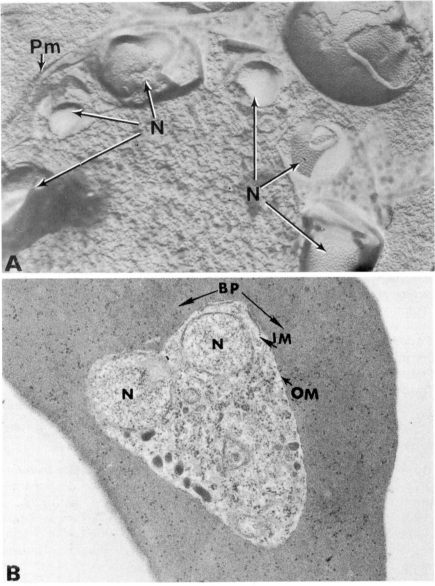

Figure 2.20 (a) Early division stage of *Plasmodium gallinaceum*. In this freeze-cleaved preparation of an infected chicken erythrocyte six nuclei can be seen lined up just under the plasma membrane (from Seed *et al.* 1971). (b) Thin section electron micrograph of *Babesia microti* in the process of reproduction. Two nuclei (N) are present, one in each bud primordium (BP). The inner (IM) and outer (OM) membranes, characteristic of the merozoites, are present in the region of the buds (from Kreier *et al.* 1975).

Merogony (= schizogony)   Merogony, which was originally regarded as a form of multiple fission, has been shown by electron microscopy to be a type of multiple budding. After two or more nuclear divisions, the "daughter" nuclei move to the periphery of the parent organism (Fig. 2.20a) and "daughter" individuals or merozoites develop as buds, one related to each nucleus (Fig. 2.20b). Eventually the merozoites (into each of which one nucleus has entered) break off; the meront is destroyed in the process, all that remains of it being a residual body of cytoplasm (and, in the case of erythrocytic meronts of *Plasmodium*, the malarial pigment). It is characteristic of merogony that bud formation does not begin until after all nuclear divisions are complete. The number of merozoites produced ranges from four to many thousands. In some very large meronts the area available for the production of merozoites is increased by complex invaginations of the surface, and this may sometimes result in the complete separation of parts of the meront, which are then called cytomeres. Merogony is found only in the Apicomplexa (and, perhaps, in the Microsporida).

It is interesting to note that, at least in the genus *Plasmodium*, the way in which the sporozoites are formed during sporogony is very similar to merogony, and probably the two processes should be regarded as the same.

Endodyogeny   This rather rare type of asexual reproduction, so far recorded only in the Toxoplasmea, consists of the development of two "daughter" individuals within a single parent which is destroyed in the process (Fig. 2.21). Although sometimes called "internal budding", endodyogeny may perhaps be regarded as a special kind of merogony.

Sexual reproduction

Among symbiotic protozoa, sexual reproduction is known to occur only in the Apicomplexa, Opalinata, Hypermastigida (Zoomastigophorea parasitic in the gut of termites and insects) and Ciliophora. The latter have a very specialized form of sexual reproduction (conjugation), which is described in Chapter 10. In the Apicomplexa, sexual reproduction consists basically of the fusion of two sexual individuals, male and female, usually anisogametes. Anisogametes are so called as they differ in size, and the male gametes bear one or more flagella (except in certain Gregarinia), while the females do not.

Sexual reproduction in the Opalinata occurs by the fusion of two multiflagellate anisogametes (which differ slightly in size). These gametes emerge from spores (cysts) which are produced only during the host's breeding season, apparently under the influence of the onset of sexuality in the host itself (which is usually an amphibian). The

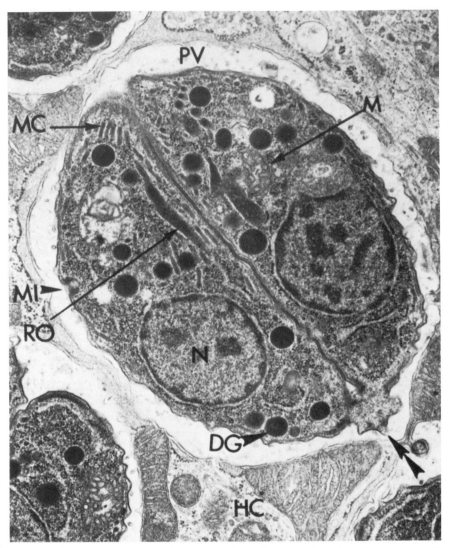

Figure 2.21 Two *Toxoplasma* daughter cells still retained within the parent cell membrane and still joined at the posterior end (arrow). Micronemes (MC), rhoptries (RO), a micropore (MI), dense granules (DG), nucleus (N), and mitochondrion (M) are fully developed in each daughter cell. The organisms are within a host cell (HC). Production of daughter cells within the parent cell is called endodyogeny (the electron micrograph was provided by H. G. Sheffield).

spores (cysts) are therefore available to infect the young tadpoles, in which fertizilization occurs.

The occurrence of sexual reproduction among the hypermastigid flagellates is controlled by the production of moulting hormone by the host. It consists of the fusion of two flagellate gametes which may be similar or dissimilar (i.e. isogamous or anisogamous), and which, in some species, are indistinguishable from the trophozoites.

Reduction division of the nucleus (meiosis) occurs at various stages in the life cycle of those protozoa which undergo sexual reproduction. In the Apicomplexa, and in some hypermastigid flagellates, it occurs at the first nuclear division of the zygote; in other hypermastigids, in the Ciliophora and in the Opalinata, it occurs during the production of the gametes (as it does in *Homo sapiens*). Thus the former organisms are haploid throughout most of their life cycle, and the latter are diploid. The process of meiosis in protozoa is, as far as is known, very similar to that in the metazoa.

FURTHER READING

Adam, K. M. G, J. Paul & V. Zaman 1971. *Medical and veterinary protozoology: an illustrated guide*. Edinburgh: Churchill Livingstone.

Aikawa, M. & T. M. Seed 1980. Morphology of plasmodia. In *Malaria*, vol. 1, J. P. Kreier (ed.), 285–344. New York: Academic Press.

Ash, L. R. & T. C. Orihel 1980. *Atlas of human parasitology*. Chicago: American Society of Clinical Pathologists.

Baker, J. R. 1982. *The biology of parasitic protozoa*. London: Edward Arnold.

Blum, J. J. & M. Hines 1979. Biophysics of flagellar motility. *Quarterly Review of Biophysics* **12**, 103–80.

van den Bossche, H. (ed.) 1972. *Comparative biochemistry of parasites*. New York: Academic Press.

Cox, F. E. G. 1985. Chromosomes of malaria parasites and trypanosomes. *Nature* **315**, 280–1.

Dogiel, V. A. 1964. *General parasitology*. Revised and enlarged by Y. I. Polyanski & E. M. Kheisin (translated by Z. Kabata). Edinburgh: Oliver & Boyd.

Gutteridge, W. E. & G. H. Coombs 1977. *Biochemistry of parasitic protozoa*. London: Macmillan and Baltimore. University Park Press.

Jahn, T. L. & E. C. Bovee 1968. Locomotion of blood protists. In *Infectious blood diseases of man and animals*. D. Weinman & M. Ristic (eds), 394–436. New York: Academic Press.

Kreier, J. P. (ed.) 1977. *Parasitic protozoa*, vols 1–4. New York: Academic Press.

Levandowsky, M. & S. H. Hutner (eds) 1979. *Biochemistry and physiology of protozoa*, 2nd edn, vols 1–4. New York: Academic Press.

Levine, N. D. 1973. *Protozoan parasites of domestic animals and of man*, 2nd edn. Minneapolis: Burgess.

Nisbet, B. 1984. *Nutrition and feeding strategies in protozoa*. London: Croom Helm.

Rudzinska, M. A. & K. Vickerman 1968. The fine structure. In *Infectious blood diseases of man and animals*, D. Weinman & M. Ristic (eds), 217–306. New York: Academic Press.

Satir, P. 1984. The generation of ciliary motion. *Journal of Protozoology* **31**, 8–12.

Sleigh, M. 1973. *The biology of protozoa*. London: Edward Arnold and New York: American Elsevier.

Sleigh, M. 1973. *Cilia and flagella*. London: Academic Press.

Society of Protozoologists 1984. Symposium – the structure and function of cilia and flagella. *Journal of Protozoology* **31**, 7–40.

Trager, W. 1967. Cultivation and nutritional requirements. In *Infectious blood diseases of man and animals*, D. Weinman & M. Ristic (eds), 149–74. New York: Academic Press.

Walliker, D. 1983. *The contribution of genetics to the study of parasitic protozoa*. Letchworth, England; Research Studies Press, and New York: Wiley.

Whitfield, P. J. 1979. *The biology of parasitism: an introduction to the study of associating organisms*. London: Edward Arnold and Baltimore: University Park Press.

Zaman, V. 1979. *Atlas of medical parasitology*. Balgowlah, Australia: Adis Press.

CHAPTER THREE

Trypanosomes and related organisms

These organisms belong to the class Zoomastigophorea, order Kineto-plastida – so named because its members possess an organelle called a kinetoplast. This is seen by light microscopy as a small, usually round or oval body which stains similarly to a nucleus, situated near the base of the flagellum. For many years its true nature puzzled protozoolo-gists, but electron microscopy revealed that the kinetoplast is an unusually large mass of mitochondrial DNA, contained within a very large mitochondrion. DNA occurs in mitochondria of many (possibly all) other organisms, but the Kinetoplastida have more of it than do most other animal cells; the significance of this is still debatable. No sexual development is known for any member of the order, although there is indirect evidence that some kind of genetic exchange may occur in at least one species, *Trypanosoma brucei*. There are two suborders within the Kinetoplastida, the Bodonina and the Trypano-somatina, the latter being the more important from the parasitologist's viewpoint.

SUBORDER 1: BODONINA

Many of these organisms are free living. They can be distinguished from the Trypanosomatina by having more than one flagellum (usually two). Most of the symbiotic forms live in the intestines of vertebrates and invertebrates, but some species of the genus *Cryptobia* inhabit the blood of fish and these are often called trypanoplasms (Fig. 3.1) since they were once grouped in a separate genus, *Trypanoplasma*. They are not usually pathogenic. Trypanoplasms are transmitted by leeches, in which they multiply in the gut. Eventually they migrate forwards to the proboscis sheath, and some are injected into the blood of the next fish on which the leech feeds.

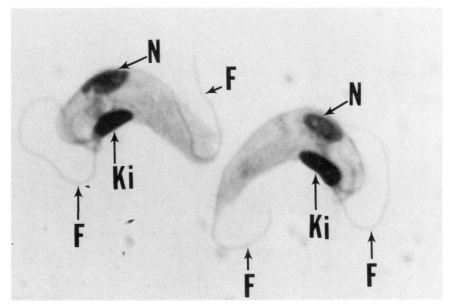

Figure 3.1 *Cryptobia salmonsitica*, a trypanoplasm from a salmonid fish. A prominent kinetoplast (Ki) and two flagella (F) as well as the nucleus (N) can be seen in each of the two organisms in this photomicrograph (provided by Professor P. T. K. Woo, University of Guelph, Ontario, Canada).

SUBORDER 2:
TRYPANOSOMATINA

All members of this order are symbiotic and several are important pathogens. All are elongate, slender protozoa at least at some stage of their life cycles, with a single nucleus and a kinetoplast situated near the origin of the single anterior flagellum by means of which they swim actively. Different forms are recognized, depending upon the position in the body of the kinetoplast and basal body, and the course taken by the flagellum. Some genera exist for part of their life cycle as nonflagellate or amastigote individuals. The genera *Leptomonas, Herpetomonas, Crithidia* (Fig. 3.2), *Blastocrithidia*, and *Rhynchoidomonas* are exclusively symbionts of insects (and a few other invertebrates); they all inhabit the gut, and are presumably transmitted via the feces, sometimes at least as encysted amastigote forms. The genus *Phytomonas* (Fig. 3.3) inhabits plants, especially succulents, and is transmitted by Hemiptera which feed on the plant juices. *Phytomonas* has been recorded from various parts of the world, usually the warmer regions but by no means always in the tropics; morphologically, it resembles *Leptomonas*.

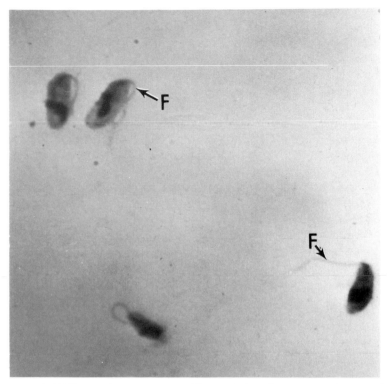

Figure 3.2 *Crithidia* sp. from an insect. The organisms are promastigote with the flagellum (F) emerging at the anterior end of the body.

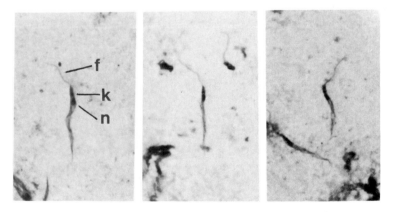

Figure 3.3 *Phytomonas* sp. from the sap of a *Euphorbia* plant in Shinyanga, Tanzania. The organisms are promastigotes, and the anterior flagellum (f), kinetoplast (k) and, just behind it, the nucleus (n) are visible.

Three genera are symbionts of vertebrates: *Leishmania, Trypanosoma,* and *Endotrypanum*. Almost all species of these three genera also inhabit blood-sucking invertebrates (usually insects or leeches), by means of which they are transmitted from one vertebrate to another. Some of them are pathogenic to their vertebrate hosts but, with one possible exception (*T. rangeli*), there is no evidence that any harms its invertebrate host. *Endotrypanum* (Fig. 3.4) is unique among the Trypanosomatidae in that it lives inside the erythrocytes of its hosts (sloths of South and Central America); the parasites are either epi- or promastigote and are perhaps transmitted by sandflies. The full life cycle is unknown.

Leishmania exists in two forms: amastigote (i.e. rounded, nonflagellate) individuals (Fig. 3.5) in its vertebrate host and elongated, flagellate organisms in the invertebrate. The flagellate form is of the type known as promastigote (Fig. 3.6). In the promastigote the basal body and kinetoplast are close to the anterior end, and the flagellum emerges through a short intucking of the body surface, the "reservoir"

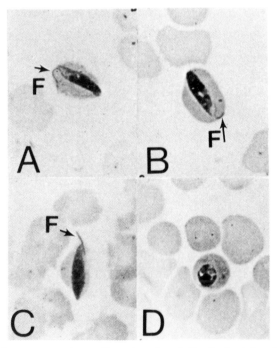

Figure 3.4 *Endotrypanum* sp. in erythrocytes of a sloth. A short free flagellum (F) is visible in some of the specimens (a, b and c). The organisms may be elongated and distort the erythrocyte (c) or be rounded and cause little or no distortion (d) (from Shaw 1969).

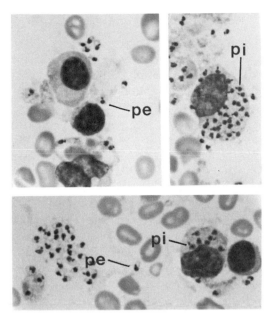

Figure 3.5 *Leishmania* parasites in a smear made from the spleen of an infected hamster; both intracellular (pi) and extracellular (pe) parasites can be seen, the latter having been released from cells ruptured during the making of the smear.

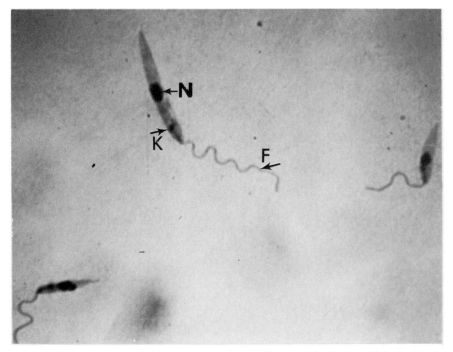

Figure 3.6 *Leishmania tropica* promastigote forms. These are the forms which develop in the insect vector and also in culture. The anterior flagellum (F), kinetoplast (K) and nucleus (N) can be seen in this photomicrograph of stained organisms.

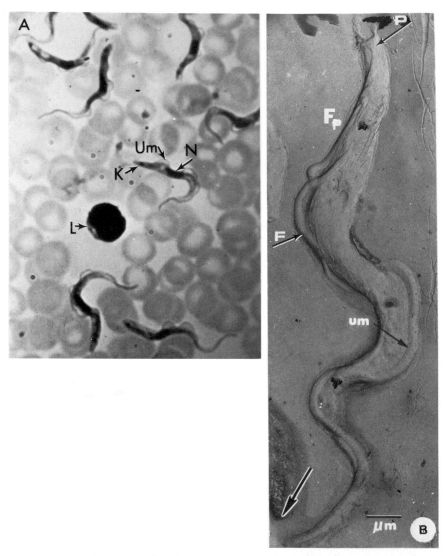

Figure 3.7 (a) *Trypanosoma brucei* trypomastigotes. A number of trypomastigotes are visible in this Giemsa-stained thin blood film. The undulating membrane (Um), nucleus (N) and small, round, posteriorly situated kinetoplast (K) labelled on the organism in the center of the field are visible in most of the organisms. The round, intensely stained structure in the center of the field is a lymphocyte (L). (b) In this carbon replica of a trypanosome the flagellum (F) can be seen to emerge from the pocket (Fp) near the posterior end (P) of the body and wind around the body, to which it is tightly attached and from which it may draw out the undulating membrane (Um) (from Seed *et al.* 1972).

or "flagellar pocket" (see Fig. 2.10). (*Leptomonas* and *Phytomonas* also exist in these two forms.)

Organisms of the genus *Trypanosoma* are characterized by the fact that, at some stage of their life history, all occur as trypomastigote forms (Fig. 3.7) – slender, flagellate individuals in which the kinetoplast and basal body are near the posterior end, and the flagellum emerges through a short "pocket", running anterolaterally. The flagellum then passes forwards along the surface of the organism's body, to which it is attached by hemidesmosomes in such a way that, as the flagellum undulates, it draws out a fin-like expansion of the body to form the undulating membrane. Some species of *Trypanosoma* also exist at different stages of their life cycle as amastigotes and promastigotes, as well as in a fourth form, the epimastigote (Fig. 3.8). In this last form, the kinetoplast and basal body lie close to, usually beside, the nucleus; the flagellum emerges, and is attached to the body to form an undulating membrane in the same way as in the trypomastigote form. Some species of *Trypanosoma* may lack the amastigote and promastigote stages, while others exist only as trypomastigotes. The amastigote–promastigote type may represent the ancestral form of the subfamily, epi- and trypomastigote stages being subsequent evolutionary additions; thus the life cycle of *Trypanosoma* can be regarded as an ontogenic recapitulation of phylogeny, from which some of the more recently evolved species have omitted one or more of the "ancestral" stages.

GENUS *LEISHMANIA* ROSS, 1903

These are Kinetoplastida which inhabit macrophages of vertebrates (mammals and reptiles) and the gut of sandflies (Diptera, Psychodidae,

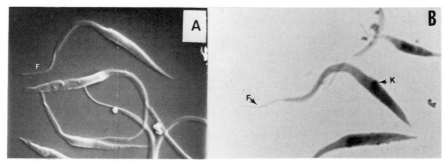

Figure 3.8 *Trypanosoma cruzi* epimastigotes. (a) Scanning electron micrograph, (b) photomicrograph of a Giemsa-stained preparation. The kinetoplast (K) lies near, but anterior to the nucleus. The flagellum (F) extends beyond the body.

Phlebotominae); they exist as amastigote forms in the former, and as promastigotes in the latter.

Morphology and life cycle

The morphology and life cycle are similar in those species for which they are known. In the vertebrate host the organisms exist solely as

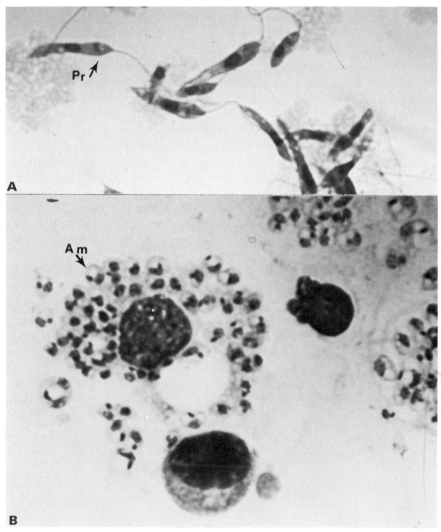

Figure 3.9 *Leishmania* in a cell culture. (a) Promastigotes (Pr) just after addition to a culture of macrophages, (b) amastigotes (Am) filling the macrophage cytoplasm after incubation.

Figure 3.10 *Phlebotomus* feeding. Only the female feeds on blood. The protein in the blood is used for egg production. The leishmanial promastigotes are introduced into the wound with the saliva at the time of feeding.

amastigote forms, usually 2–4 µm in diameter, often called "Leishman–Donovan" or "LD" bodies after the names of their first observers (Fig. 3.5). These are ingested by the macrophages as part of the latter's phagocytic activity, but instead of being destroyed, the protozoa apparently resist the lysosomal enzymes released into the phagolysome; they then multiply by binary fission within the macrophages (Fig. 3.9). When an infected macrophage dies, the liberated protozoa are presumably ingested by other macrophages. Thus more cells become infected. Infected macrophages in the blood or skin are ingested by sandflies (Phlebotominae) when the insects are feeding (Fig. 3.10) (only the females feed on blood). In the midgut of the

51

sandfly, the protozoa emerge from the macrophages, transform into promastigote forms by elongating and developing a flagellum, and commence multiplying by binary fission. Three types of development are recognized, depending on the location of the promastigotes within the sandfly's gut: peripylarian, hypopylarian, and epipylarian. As development proceeds the flagellates spread forwards through the insect's gut, many of them becoming attached to the mucosa by their flagella. They become established in the foregut, and by their multiplication eventually block the cavity of the proventriculus, pharynx and proboscis. At this stage, about ten days after it first ingested protozoa, when the insect again attempts to feed (which it usually does about every five days), its efforts to pump saliva down the blocked proboscis and to draw up blood result in the injection of many flagellates into the host's skin. Here they are phagocytosed by macrophages and, after reverting to the amastigote form, they commence dividing. Transmission may also result from the crushing of infected sandflies while they are feeding, their gut contents entering the vertebrate through the puncture made by the proboscis. Blood transfusion from an infected donor has also transmitted the infection.

Important species

Most species of *Leishmania* are grouped into one of three "species complexes", the constituents of which can be regarded as subspecies (Table 3.1).

The *L. donovani* complex   The members of this group cause visceral leishmaniasis or kala azar of man. They occur throughout the warmer parts of Asia, the Mediterranean coasts, North and East Africa and (as *L. d. chagasi*) in South America. In the Mediterranean region, China, central Asia and South America, dogs are often infected and serve as reservoirs of the infection; in India man is the only known vertebrate host. Infected macrophages occur in all tissues (Fig. 3.11), including the blood; the progress of the disease is slow but, unless treated, the disease is usually fatal.

The *L. mexicana* complex   The members of this group cause New World cutaneous leishmaniasis (a similar condition produced in the Old World by *L. tropica* is shown in Fig. 3.12). They occur in Central and South America, Mexico and (rarely) Texas. Various forest-dwelling mammals may be reservoir hosts. Amastigotes are found in ulcers in the skin, usually only one or a few on each patient. The disease is normally mild, usually ending spontaneously. However, amastigotes of *L. m. amazonensis* may, and those of *L. m. pifanoi* always, spread

Table 3.1 Species of *Leishmania* known to infect man

Species	Subspecies	Geographical area	Reservoir host(s)	Diseases produced
L. donovani	*L. d. donovani*	Asia, (including China, USSR, India), Africa, southern Europe	usually dogs and wild canids; rarely rodents or (in Sudan) felids; none known in India	Old World visceral leishmaniasis (kala azar); post-kala azar dermal leishmanoid
	*L. d. infantum**			
	*L. d. chagasi**	Mexico, Central and South America	dogs, foxes	New World visceral leishmaniasis
L. mexicana	*L. m. mexicana*	USA (Texas), Mexico, Central and South America	usually rodents	New World cutaneous leishmaniasis
	L. m. garnhami			
	L. m. venezuelensis			
	*L. m. amazonensis**†			
	L. m. pifanoi†			
L. braziliensis	*L. b. braziliensis*	Central and South America	rodents, sloths; rarely primates and procyonids	New World cutaneous leishmaniasis; muco-cutaneous leishmaniasis
	*L. b. guyanensis**			
	*L. b. panamensis**			
L. tropica		Asia (including USSR, India), Africa, southern Europe	dogs, rodents	Old World cutaneous leishmaniasis
L. major				
L. aethiopica†		Africa (east and north-east)	hyraxes	
L. peruviana		Peru (highlands)	dogs	Uta

* These subspecies are sometimes treated as distinct species.
† May cause diffuse cutaneous leishmaniasis (DCL).

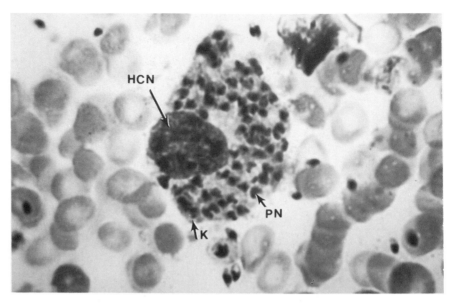

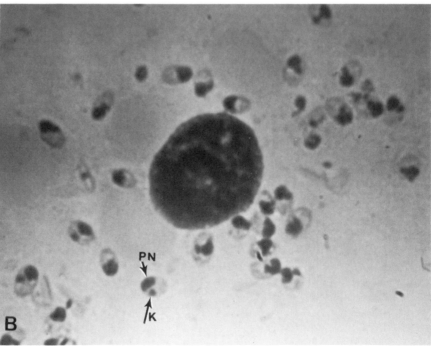

Figure 3.11 These photomicrographs are of Giemsa-stained preparations of *Leishmania donovani*. In (a) an intact macrophage with its cytoplasm full of amastigotes is present. The host cell nucleus (HCN), parasite nuclei (PN) and kinetoplasts (K) are visible in the host cell cytoplasm. In (b) the host cell has broken, probably in preparation of the film, and the amastigotes are dispersed. Parasite nuclei (PN) and kinetoplasts (K) are discernible in many of the oval amastigotes.

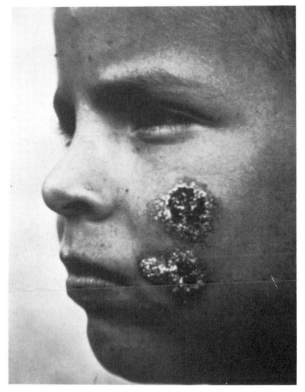

Figure 3.12 *Leishmania tropica*, tropical sore. In this photograph, several lesions are seen on the boy's cheek. This is a common site of infection only because it is a common site for the insect vector to bite. As the infection is self-limiting and results in immunity, it is a common practice to induce a lesion in an inconspicuous area by inoculation with parasites to prevent later infection and scarring of the face.

through the skin to cause an extremely disfiguring condition called diffuse cutaneous leishmaniasis (DCL); this condition, also caused by *L. aethiopica*, is illustrated in Fig. 3.13.

The *L. braziliensis* complex   The members of this group occur only in Central and South America. *L. b. braziliensis* itself, especially, may cause a very severe human disease – mucocutaneous leishmaniasis or espundia (Fig. 3.14). The mucous membranes of the nose, mouth and pharynx become infected and ultimately destroyed. Spontaneous recovery is rare.

Another, looser grouping of species, not now considered to be a homogeneous complex, is that of the species causing Old World cutaneous leishmaniasis, *L. tropica* (Fig. 3.12), *L. major* and *L. aethiopica*.

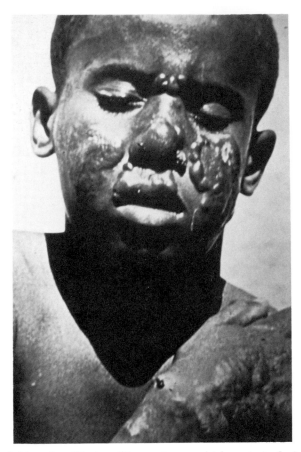

Figure 3.13 *Leishmania aethiopica*, diffuse cutaneous leishmaniasis. In individuals who lack a strong cellular immune response to the infection, the organisms may spread in the subcutaneous tissue causing extensive swelling such as is seen on the face and hand of this individual.

Lesions are normally restricted to the skin, and the disease is usually self-limiting; however, a chronic condition may develop and *L. aethiopica*, like some of the New World species, may cause DCL (Fig. 3.13). This condition, seems, at least when caused by *L. aethiopica*, to result from cellular immunodeficiency in the patient.

All species of *Leishmania* which infect man (excepting *L. donovani* in India) are zoonoses and have vertebrate "reservoir" hosts other than man.

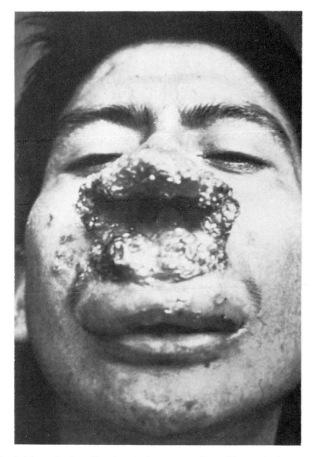

Figure 3.14 *Leishmania braziliensis* causes espundia. The infection may be very destructive and usually develops at a mucocutaneous junction (e.g. lips or nose); if untreated, it may progressively erode the underlying tissue. An extremely destructive lesion in the nasal region is illustrated in this figure.

Diagnosis

Human leishmaniasis may be diagnosed by examining smears of material obtained by puncturing or scraping suspected lesions; the smears are stained with Giemsa's stain in the same way as a thin blood film (see Ch. 11) and parasites, if present, can be seen. A more sensitive method is to inoculate material obtained as described above into laboratory animals (particularly hamsters) or into blood–agar cultures (Ch. 11); if cultures are used, the material from the lesions must be obtained aseptically.

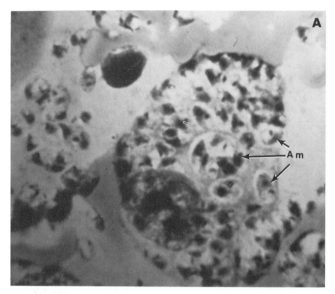

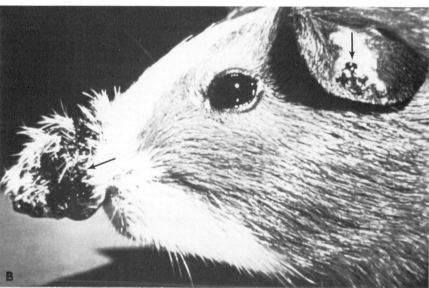

Figure 3.15 *Leishmania enriettii* is a parasite of the guinea-pig in which it causes cutaneous leishmaniasis. (a) Photomicrograph of amastigotes (Am) in exudate cells from a lesion. The amastigotes of this organism are larger than those of other species of *Leishmania*. (b) This guinea-pig has a very prominent lesion (arrow) on its nose and also a smaller lesion (arrow) on its ear.

Other species

Apart from infections of *L. donovani*, *L. tropica* and *L. peruviana* in dogs, and the first two sometimes in cats, there is no leishmaniasis of domestic animals. Wild rodents and lizards may be infected with species of *Leishmania* which are morphologically similar to those of man. *L. enriettii*, a parasite of South American guinea-pigs, has unusually large amastigotes – about 3×6 µm (Fig. 3.15)

GENUS *TRYPANOSOMA* GRUBY, 1843

The trypanosomes are Kinetoplastida inhabiting the blood and sometimes other tissues of vertebrates and, usually, the gut of blood-sucking invertebrates. They exist as trypomastigote forms for at least part of their life cycle in both hosts.

Trypanosomes of non-mammals

Species of *Trypanosoma* are common symbionts of fishes (fresh-water and marine), amphibia, reptiles and birds throughout the world. Usually these species are relatively large organisms (about 50–100 µm long) and are very scanty in the blood of their hosts. All, as far as is known, are nonpathogenic. The vectors of the trypanosomes of terrestrial vertebrates are blood-sucking arthropods (usually insects, occasionally mites); the trypanosomes of aquatic forms are transmitted by leeches (Annelida, Hirudinea). In the vector's intestines they multiply first as epimastigote forms and then change into small trypomastigote forms ("metacyclic forms" or metatrypanosomes). The latter are capable of reinfecting the vertebrate host; they usually develop in the vector's hindgut and are introduced to the vertebrate either by fecal contamination of the wound produced by the vector when it feeds, or by the vertebrate crushing infected vectors in its mouth. The species transmitted by leeches, however, often complete their development in the vector's foregut and so are injected with the leech's saliva into the next invertebrate on which it feeds. Much remains unknown about this group of trypanosomes. One symbiont of frogs, *T. rotatorium*, appears normal in the tadpole but, after the latter's metamorphosis into an adult frog, becomes enlarged and "leaf-like" in shape.

Trypanosomes of mammals

All the important pathogens are included in this group, which has been divided by Hoare into two sections (Stercoraria and Salivaria),

Table 3.2 Main morphological features of some mammalian trypanosomes as seen in blood or tissue fluids of the vertebrate host.

Subgenus	Species	Average length (µm)	Average breadth (µm)	Kinetoplast size	Kinetoplast position	Posterior end	Free flagellum	Position of nucleus
Megatrypanum	T. theileri	60–100	2–3	medium	not terminal	pointed (long)	long	central
	T. melophagium	40–60	2–3					
Herpetosoma	T. rangeli	25–35	2	medium	not terminal	pointed (long)	long	slightly anterior
	T. lewisi	21–36	2	large, rod-like	not terminal	pointed (long)	long	slightly anterior
Schizotrypanum	T. cruzi	15–24	1.5	large	subterminal	pointed (short)	long	central
Duttonella	T. vivax vivax		1.5–2	large	terminal	blunt, rounded	long	central
	T. v. viennei		1.5					
	T. uniforme	14–17	1.5					
Nannomonas	T. congolense	12–18	1–2	medium	subterminal, marginal	usually blunt	usually absent	central
	T. simiae	17–18	1–2					
Pycnomonas	T. suis	15	3.5	small	subterminal	pointed (very short)	short	central
Trypanozoon	T. brucei (sensu lato)							
	Slender forms	30	1.5			blunt, often truncate	long	central
	Stumpy forms	18	3.5	small	subterminal	blunt, rounded	usually absent	central (except in posteronuclear form)
	T. evansi evansi T. e. equiperdum	20–28	1.5	small	subterminal	blunt or truncate	long	central
	T. equinum	20–28	1.5	absent		blunt or truncate	long	central

each containing several subgenera. The chief morphological characteristics of the more important species of both sections are summarized in
Table 3.2, and a simplified "key" for their identification in stained
blood films is given in Table 3.3.

Table 3.3   A simplified key to the more important mammalian
trypanosomes in stained thin blood films.*

1	Trypanosomes more than 40 μm long	*T. theileri*, etc.
	Trypanosomes less than 40 μm long	2
2	All individuals with free flagellum	3
	Some or all individuals without free flagellum	11
3	Kinetoplast not terminal, medium-sized to small (or absent)	4
	Kinetoplast terminal, large	9
4	Posterior end sharply pointed	5
	Posterior end bluntly pointed or rounded	10
5	Kinetoplast medium-sized or large	6
	Kinetoplast small	8
6	Trypanosomes curved (C-shaped); kinetoplast large, round	*T. cruzi*
	Not as above	7
7	Kinetoplast medium-sized, round	*T. rangeli*
	Kinetoplast large, rod-shaped	*T. lewisi*
8	Long, slender trypanosomes (rare Central African species)	*T. suis*
9	Trypanosomes more than 20 μm long	*T. vivax*
	Trypanosomes less than 18 μm long	*T. uniforme*
10	Kinetoplast present, small	*T. evansi, T. equiperdum* and old laboratory strains of *T. brucei (sensu lato)*
	Kinetoplast absent	*T. equinum*
11	All individuals without free flagellum	*T. congolense*
	Some individuals have, or appear to have, a free flagellum	12
12	Differentiated into long, slender forms with free flagellum, short stumpy forms without free flagellum, and intermediates; kinetoplast not markedly marginal in position	*T. brucei (sensu lato)*
	Not as above; kinetoplast marginal	*T. simiae*

*In practice, the species of animal and the geographical location from which the
trypanosomes were isolated are usually valuable aids to identification.
This key gives only certain differential characters of the groups concerned: it is not a full
description of them, nor does it represent their phylogenetic relationships.

SECTION A: STERCORARIA

All members of this section have two hosts, a mammal and an insect vector. The insect ingests trypomastigotes when it feeds on the blood of an infected mammal; these forms quickly change into epimastigotes, chiefly by a forward movement of the kinetoplast and flagellar basal body (or kinetosome). The epimastigotes then divide repeatedly by longitudinal binary fission in the insect's midgut and gradually migrate backwards to colonize the hindgut. Finally, metacyclic trypanosomes (= metatrypanosomes) develop in the hindgut. The metatrypanosomes probably do not divide until after they have entered the mammalian host. The fact that they develop in the hindgut of the vector is one of the main features separating the Stercoraria from the Salivaria. *Trypanosoma rangeli*, in other respects a stercorarian, may develop metatrypanosomes in the salivary glands of its vector. Metatrypanosomes may be passed out in the vector insect's feces, or they may be liberated from an infected insect if the latter is crushed by the mammal on which it is feeding. They can then enter the mammal's bloodstream in various ways: they may penetrate the puncture produced by the insect's proboscis; they may be rubbed into lesions produced by scratching the skin in response to the irritation of the insect's bite; they may penetrate the mucous membrane of the mouth if the mammal either licks up insect feces or crushes whole insects in its mouth during its cleaning operations. Trypanosomes can probably penetrate an intact mucous membrane, though this is not certain; they may depend on the presence of small lesions, as they do to pass through skin. All these methods are examples of contaminative transmission, another distinctive feature of the Stercoraria. Once in the vertebrate host, the trypanosomes continue their life cycle, usually extracellularly (though intracellularly in one important subgenus, *Schizotrypanum*), by undergoing binary fission. It is a third characteristic of the Stercoraria that multiplication in the vertebrate host occurs only at certain phases of the life cycle, and that the individuals which divide are not trypomastigotes. Trypomastigotes develop from the products of these periodic bouts of multiplication.

There are three subgenera in this section: *Megatrypanum*, *Herpetosoma* and *Schizotrypanum*.

Subgenus Megatrypanum

These are nonpathogenic species infecting ruminants; none can infect man. The trypomastigote forms in the vertebrate host are large (50 μm or more in length), have a long pointed posterior end extending well behind the kinetoplast, and a free flagellum (i.e. the flagellum

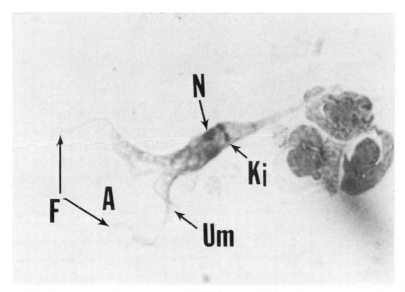

Figure 3.16 *Trypanosoma (Megatrypanum) theileri,* a nonpathogenic trypanosome found in domestic cattle throughout the world. The organism pictured is in the process of division. Division has started at the anterior (free flagellum bearing) end (A). The flagellum (F) extends along the body forming the undulating membrane (Um). The nucleus (N) and kinetoplast (Ki) are already duplicated in the portion of the body which has not split. The photomicrograph is of a fixed and stained preparation made from an infected chick embryo (photomicrograph provided by Professor P. T. K. Woo, University of Guelph, Ontario, Canada).

extends beyond the anterior end of the cell body). Species include *T. (Megatrypanum) theileri* of cattle (Fig. 3.16), transmitted by horse-flies (Diptera, Tabanidae: see Fig. 3.23B) and *T. (M.) melophagium* of sheep, transmitted by louse-flies (Diptera, Hippoboscidae), both of which occur throughout the world. They are usually present in the blood of their vertebrate hosts in very small numbers, which makes them very difficult to find in blood films. Division of *T. (M.) melophagium* in the vertebrate host has not been seen.

Subgenus Herpetosoma

Most species of this subgenus are nonpathogenic symbionts of rodents. The trypomastigote forms in the blood of the vertebrate host are similar in all these species, being about 30 μm long, with a pointed posterior end extending some distance behind the kinetoplast, and a free flagellum. The nucleus is often noticeably in front of the mid-point of the body. Most species are very host-specific, being unable to infect even closely related vertebrate species: e.g. *T. lewisi* of the rat cannot infect mice, and *T. musculi* of mice cannot infect rats, although morphologically the two parasites are indistinguishable.

Important species

*Trypanosoma (Herpetosoma) rangeli** (Figure 3.17)   This species infects man as well as dogs, cats, opossums, and monkeys in parts of South America. Though morphologically resembling others in this section, it is exceptional in that its development in its vector, the bug *Rhodnius prolixus* (Hemiptera, Reduviidae), takes place in the hemocoel (after an initial phase in the midgut), and metacyclic trypanosomes develop in the bug's salivary glands. After development in the salivary glands the metacyclic trypanosomes are inoculated into the next vertebrate on which the bug feeds. This type of inoculative transmission is characteristic of the second section of the trypanosomes of mammals, the Salivaria. Sometimes transmission occurs via the hindgut by contaminative methods also. It is not certainly known whether *T. rangeli* divides at all in its vertebrate host, or, if it does, where and in what form division occurs. *Trypanosoma rangeli* seems to be non-pathogenic to its vertebrate hosts, though there is some evidence that it is harmful to its invertebrate host, apparently unique behavior for a trypanosome.

Trypanosoma H. (lewisi) (Figure. 3.18)   This is a cosmopolitan non-pathogenic parasite of the wild rat (*Rattus norvegicus* and *R. rattus*). Its vectors are the rat fleas *Xenopsylla cheopis* and *Ceratophyllus fasciatus* (Siphonaptera), in which it develops in the mid- and hindgut. Reports are conflicting, but at least in some circumstances it has an intracellular phase of development in a midgut cell. After introduction (by contaminative methods) of the metacyclic forms into the rat, division (multiple fission of epimastigote forms) occurs in the blood capillaries of the viscera, especially the kidneys, for 4–5 days before flagellates are seen in the blood. At first these are of various shapes and sizes, but about 7–10 days after its infection the rat produces an antibody (ablastin) which inhibits further division and also probably selectively kills the dividing forms. Thereafter only "adult" trypomastigote stages are seen in the blood, until eventually (usually one month later) the rat secretes a second antibody which kills all the trypanosomes. Such a cured rat can, sometimes at least, be reinfected, but after two or three reinfections a complete immunity develops which apparently lasts for the remainder of the animal's life. (This is what happens in the laboratory rat. The infection in wild rats has been less carefully studied.) This trypanosome is often maintained in laboratories for teaching or research purposes. It cannot infect man.

---

*Now placed in a new subgenus, *Tejeraia*.

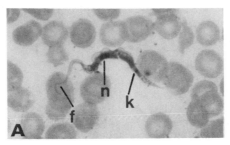

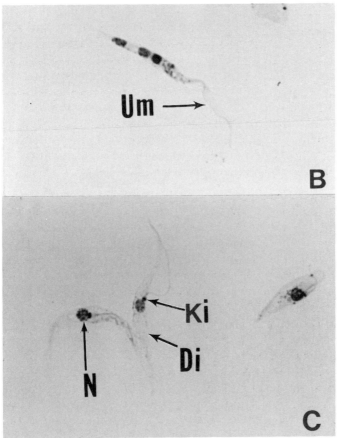

Figure 3.17 *Trypanosoma (H.) rangeli*: (a) Trypomastigote in the blood of a squirrel monkey; the flagellum (F) and nucleus (N) can be seen, as well as the small kinetoplast (K), situated some distance from the pointed posterior end: the small size and location of the kinetoplast enable *T. rangeli* to be differentiated from *T. cruzi* in stained blood films. (b, c) Epimastigotes from a culture; round kinetoplasts are visible just anterior to the nuclei (N), and the undulating membrane (Um) is prominent. Division (Di), starting at the anterior end, is well advanced in one specimen in (c).

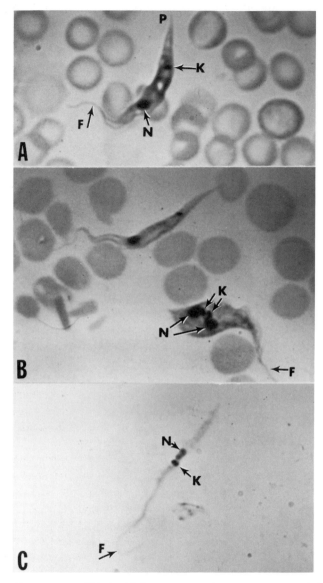

Figure 3.18 *Trypanosoma (H.) lewisi* is a stercorarian trypanosome infecting rats. Trypomastigotes occur in the blood (a) as do dividing forms (b). In cultures (and in the flea vector) epimastigotes occur (c). Flagella (F), nuclei (N), and kinetoplasts (K), are readily seen in these stained parasites. The kinetoplast is posterior to the nucleus in the trypomastigote in the blood, and just anterior to the nucleus in the dividing form in the blood and in the culture form. The blood form trypomastigotes have a long tapered posterior end (P).

Other species include *T. (H.) musculi* of mice, *T. (H.) nabiasi* of rabbits, *T. (H.) microti* of voles (all cosmopolitan), and *T. (H.) primatum* of the African chimpanzee.

Subgenus Schizotrypanum

The only medically important species of this subgenus is *T. (S.) cruzi*, a common parasite of man in South and Central America. There are many reports of similar organisms from various mammals (opossums, armadillos, rodents, monkeys) in South and Central America and the southern USA; whether all of these are identical with *T. cruzi* is uncertain at present.

At least six other species or subspecies in this subgenus have been identified in insectivorous bats (Microchiroptera) from South and North America, and Europe. As far as is known, these are unable to infect other mammals. Perhaps the subgenus is essentially a specialized group of bat parasites, of which one species (*T. cruzi*) is a "rogue" which has broken free of this rigid host-restriction, becoming pathogenic to many of its "new" hosts (including man).

Whatever its origins *T. (Schizotrypanum) cruzi* infects man all too commonly throughout South America (especially Brazil and Argentina). The disease of man caused by this organism is called Chagas disease in honor of Carlos Chagas who first described it. In Central America and Mexico, Chagas disease is not common, and in the southern USA only two human infections have been reported. However, in all these areas, and as far north as Maryland, what is almost certainly the same parasite has been recorded from various wild animals, chiefly rodents and racoons in USA, vampire bats in Panama, and armadillos, opossums and wild guinea-pigs in South America.

The human disease (Chagas disease) is really one of poverty – being associated with squalid living conditions in bug-infested mud-walled huts. The vectors are bugs, *Rhodnius prolixus* (Fig. 3.19) and others (Hemiptera, Reduviidae), which become infected by feeding on vertebrates (human or other) with parasites in their blood. Dogs and cats are important sources of human infection. Development in the vector is typical of the Stercoraria, the metacyclic trypanosomes being passed out in the bug's feces and entering the vertebrate by one or other of the contaminative routes outlined above. The bugs feed at night, and sleepy children, disturbed by the insects, often scratch the region of the bite, thus collecting infective feces on their fingers, and then rub their eyes, so introducing metacyclic trypanosomes via their conjunctiva (Fig. 3.20). In young children the disease may be rapidly fatal, but more often it is chronic; the parasites multiply in the vertebrate as large amastigote forms (about 4 μm in diameter) in

Figure 3.19 *Rhodnius prolixus* is one of the vectors of *Trypanosoma cruzi*, the cause of Chagas disease. Infection is by contamination. The bug passes feces containing infective metacyclic trypanosomes as it feeds. These can enter the bite wound or can infect if they are introduced into the eye.

various cells – macrophages, cells of the central nervous system, and also muscle cells, including those of the heart (Fig. 3.21). Thus chronic sufferers from Chagas disease may die from heart failure. The amastigotes elongate, develop a flagellum, and metamorphose into trypomastigote forms (Fig. 3.22). These enter the blood, in which they circulate for a while. They are rather small (about 20 μm long) with an acutely pointed posterior end, very large kinetoplast, and free flagellum. On dried blood films they often lie curved in the shape of a letter "C". After a time, they re-enter suitable cells, become amastigotes and commence dividing once more.

SECTION B: SALIVARIA

Almost all of the species forming this group are transmitted by tsetse flies (*Glossina* spp.); the few exceptions have evolved from forms which had a tsetse vector (Fig. 3.23A). They form a fairly closely knit group, differing in several respects from members of the Stercoraria. The vector insects, tsetse flies, are large Diptera, both sexes of which feed only on blood. The female deposits larvae singly about every ten days during her life, and the larvae pupate immediately. Tsetse flies are

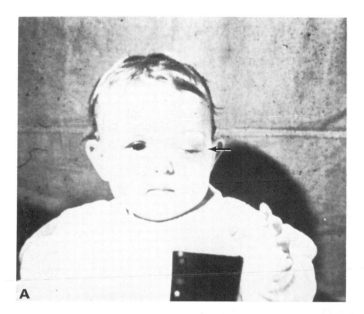

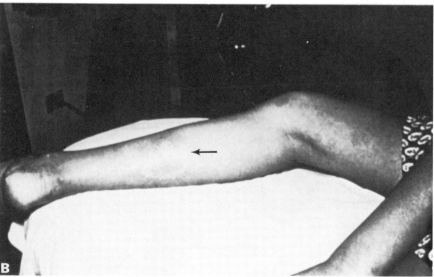

Figure 3.20 Romaña's sign (a) and chagoma (b). Chagas disease may be initiated by the introduction of feces from an infected vector into the eye. The local infection in the eye (arrow) causes inflammation with lacrimation, swelling and redness (a) and is called Romaña's sign after the individual who first described it. When the parasites gain entry through, for example, the bite wound in the skin, the local inflammation produced (here on the calf of the leg, arrow) is called a chagoma (b). These primary lesions resolve when the infection becomes generalized.

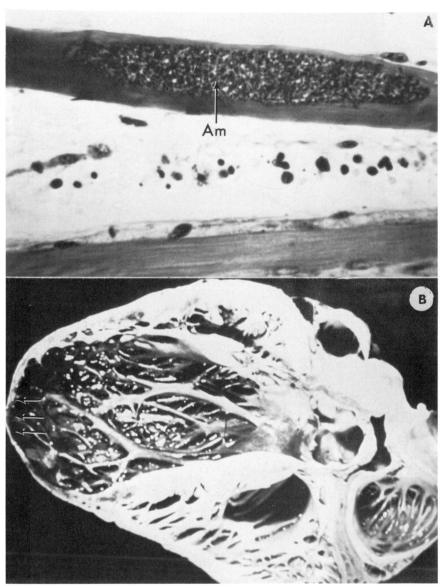

Figure 3.21 Amastigotes of *T. cruzi* in the heart. (a) Nests of amastigotes (Am) develop in the heart muscle. In heavily infected individuals, the heart muscle is damaged and as a result the heart wall is thinned and weakened. (b) The muscle in the apical region of this heart was affected and is reduced to a thin layer (arrows).

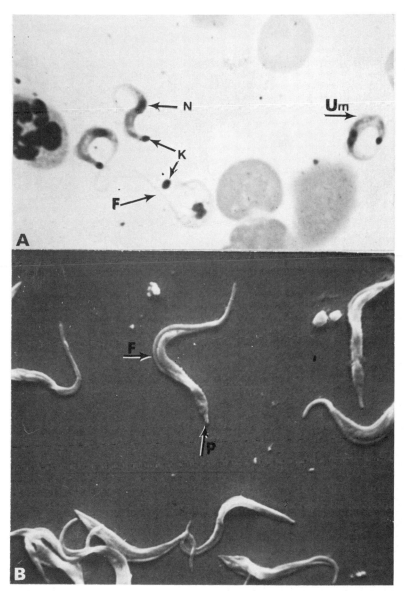

Figure 3.22 *Trypanosoma cruzi* trypomastigotes. In Giemsa-stained thin blood films (a) *T. cruzi* frequently assumes a "C" shape. The kinetoplast (K) is prominent and near the posterior end. The nucleus (N) is in the anterior portion of the body and a free flagellum (F) extends from the anterior end. An undulating membrane (Um) is often visible but is not prominent. In scanning electron micrographs (b) the attachment of the flagellum (F) to the body can be readily seen, as can the sharply pointed posterior end (P).

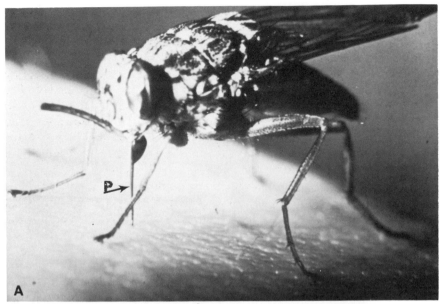

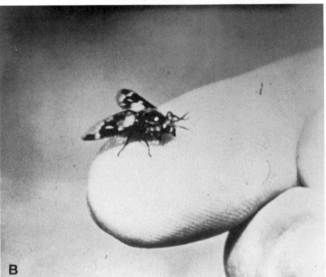

Figure 3.23 Flies of the genus *Glossina* (a) are the biological vectors of African trypanosomiasis. The infective forms are injected with the saliva when the fly feeds using its long thin proboscis (P). Tabanid flies (b), on the other hand, may spread infection mechanically if they move from one animal to another carrying blood in their probosces, but do not support growth of the trypanosomes.

found only in tropical Africa. In the vector, the trypanosomes undergo an initial period of multiplication in the trypomastigote form before changing into the epimastigote form; the developmental cycle ends with the production of metacyclic trypomastigote forms in the foregut (salivary glands or proboscis) of the tsetse fly, where they are well placed to be inoculated into the blood of the next host on which the fly feeds. A fly, once infected, remains so for the remainder of its life. The trypomastigote forms in the blood of the vertebrate host have a less elongated posterior end than do the Stercoraria; multiplication is by binary fission of the trypomastigotes in the circulating blood. Most species in this section are restricted to Africa where the group as a whole must have evolved; a few species have spread beyond this continent, and are no longer transmitted by tsetse flies. Most members of this section are important pathogens, either to man or to domestic animals. Several subgenera were defined by Hoare (Table 3.2). The geographical distribution and hosts of the various species are shown in Table 3.4.

Subgenus Duttonella

The trypomastigote forms in the vertebrate host typically have a rather broad, rounded posterior end, giving them a "club-shaped" appearance, with a large, round kinetoplast which is usually terminal in position; a free flagellum is present. In the tsetse fly, development (which follows the course outlined above) occurs exclusively in the insect's proboscis. In the laboratory, fairly high infection rates can be obtained in insects fed on infected vertebrates – up to about 50 or 60%. There are two species, differentiated by their size.

Trypanosoma (Duttonella) vivax The forms in the blood of the vertebrate host measure from 20 to 26 μm in length, with an average of more than 20 μm (Fig. 3.24). This is a common and pathogenic parasite of domestic cattle and sheep throughout Africa, between the southern limit of the Sahara and the Tropic of Capricorn. It is also found in a variety of wild animals. *Trypanosoma (D.) vivax* has also spread to Mauritius, the West Indies and parts of South America, presumably being carried there in domestic animals. In these regions it is transmitted by biting flies (Diptera), chiefly of the genera *Tabanus* and *Stomoxys*, in which it does not undergo any cyclic development, being merely carried passively in the proboscis of the insect. This type of transmission is called noncyclical (or, less appropriately, "mechanical") transmission; the insect functions only as a flying hypodermic syringe, and can remain infective only for as long as the blood in its proboscis remains moist – quite a short time. Thus noncyclical transmission depends upon a fly's being disturbed while in the act of feeding and

Table 3.4 Geographical distribution, vertebrate hosts, and transmission of salivarian trypanosomes.

Species	Geographical distribution	Main vertebrate hosts			Transmission	
		wild	domestic	laboratory	cyclical	noncyclical
T. v. vivax	West, Central, and East Africa	waterbuck, reedbuck, eland, giraffe, bushpig	cattle, sheep, goats, horses, donkeys (often pathogenic)	normally none (rats with serum supplement)	*Glossina* spp.	biting Diptera
T. v. viennei	Mauritius, South America	unknown	As for *T. v. vivax*	normally none	none	biting Diptera
T. uniforme	West Uganda, East Congo (Tanzania and Zululand rarely)	antelope	cattle, pigs (rarely if ever pathogenic)	normally none	*Glossina* spp.	biting Diptera
T. congolense	West, Central, and East Africa	many antelope, giraffe, eland, wildebeest	cattle, sheep, horses, donkeys, pigs, dogs, rarely camels (often pathogenic)	rodents (not all strains are infective; pathogenicity varies)	*Glossina* spp.	biting Diptera (less important)
T. simiae	West, Central, East, and parts of South Africa	warthog	pigs (very pathogenic)	*Cercopithecus* monkeys (very pathogenic) and splenectomized rabbits	*Glossina* spp.	biting Diptera (less important)
T. suis	Tanzania, Rwanda-Burundi, Zaire?	warthog	pigs (pathogenic to young)	none known	*Glossina* spp.	
T. b. brucei	West, Central, East, and parts of South Africa	many antelope, warthog, wildebeest	camels, horses, donkeys, dogs (fatal); cattle, pigs, sheep, goats (may be pathogenic but more chronic)	rodents, rabbits, monkeys (pathogenic)	*Glossina* spp.	biting Diptera (unimportant)

T. brucei "rhodesiense"	Central and East Africa	bushbuck, other antelope	cattle, others? MAN (pathogenic)	As for *T. b. brucei*	*Glossina* spp.	biting Diptera (unimportant)
T. b. gambiense	West and Central Africa	?	pigs and other domestic mammals, MAN (less pathogenic but more chronic than "*rhodesiense*")	as for *T. b. brucei* (less pathogenic)	*Glossina* spp.	biting Diptera (unimportant)
T. e. evansi	North Africa, Asia, South China, Philippines, Mauritius, Central and South America (eliminated from North America and Australia)	tapir (in America), deer (in Mauritius), vampire bat (South America)	camels, horses, donkeys, mules (often pathogenic); cattle, water buffalo, sheep, goats, dogs, Indian elephant	rodents (pathogenic)	none	biting Diptera
T. e. equiperdum	South America mainly, USSR, Iran and probably North and South Africa, (previously also Europe, India, North America)	none	equines (pathogenic)	rabbits (intratesticular or intrascrotal); dogs (some strains)	none	coitus
T. equinum	South America (and ?Sudan)	vampire bat and capybara (in South America)	horses (pathogenic); donkeys, mules (less pathogenic); cattle, sheep, goats, pigs (chronic)	rodents (pathogenic)	none	biting Diptera

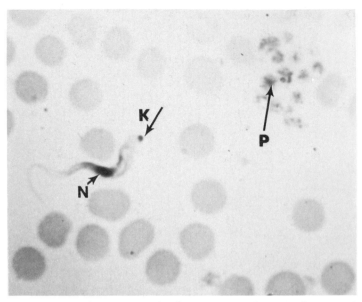

Figure 3.24 This is a photomicrograph of *Trypanosoma vivax* in a Giemsa-stained thin blood film. The nucleus (N) is centrally placed and the kinetoplast (K) is terminally located in the blunt, rounded posterior end. A cluster of platelets (P) is present in the upper right corner of the photomicrograph.

flying to another nearby animal to complete the feed. Hoare has suggested that this form of *T. vivax* should be regarded as a separate subspecies – *T. vivax viennei*.

Trypanosoma (D.) uniforme   Morphologically very similar to *T. (D.). vivax*, this species is smaller; the forms in the blood of the vertebrate measure only 12–24 µm, with a mean length of less than 18 µm. It is more restricted in its distribution than *T. vivax*, having been reported only from a few parts of tropical Africa. Its hosts are similar to those of *T. vivax*, but it seems to be rarely, if ever, pathogenic.

Subgenus Nannomonas

As seen in the vertebrate host, members of this subgenus are rather small trypomastigote forms, lacking a free flagellum (i.e. the flagellum does not extend beyond the anterior tip of the cytoplasm); however, on dried blood films there may appear to be a short free flagellum. The kinetoplast is small and usually marginal. In the tsetse fly, development begins in the midgut, and is completed in the proboscis, where the epimastigote and metacyclic trypomastigote forms are found.

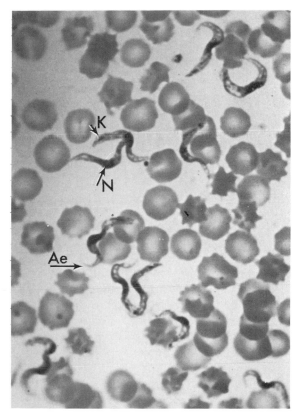

Figure 3.25 *Trypanosoma congolense*. Giemsa-stained thin blood film. *T. congolense* is monomorphic. The subterminal kinetoplast (K), centrally located nucleus (N), and anterior end (Ae) from which no free flagellum extends, are characteristics of this organism.

Infection rates in *Glossina* are not very high, even in the laboratory usually only 20–30%. The subgenus is restricted to tropical Africa.

Trypanosoma (Nannomonas) congolense (Figure 3.25)   The length of the forms seen in the blood of the vertebrate host differs considerably in different strains. The wild vertebrate hosts of *T. congolense* are mainly antelope. It is often an important pathogen of domestic animals.

Trypanosoma (N.) simiae   This species is difficult to distinguish morphologically from *T. congolense*. Generally some of the forms in the vertebrate are longer, with a better developed undulating membrane, while other individuals are indistinguishable from *T. congolense*. This parasite seems to occur only in pigs, wild and domestic; to the latter at least it is extremely pathogenic. It was named *T. simiae* because it is also very pathogenic to experimentally infected monkeys.

Subgenus Pycnomonas

This subgenus includes only one species, *T. (Pycnomonas) suis*, a rare parasite of central African wild and domestic pigs. The forms in the vertebrate are short and broad with a pointed posterior end and a short free flagellum. In *Glossina*, development begins in the midgut and ends in the salivary glands, where epimastigote and metacyclic trypomastigote forms appear.

Subgenus Trypanozoon

This subgenus seems to be in the midst of a period of rapid evolution at the present time, which makes it difficult for systematists to divide it up neatly into distinct species. The "classical" view is that there are six species in the subgenus; some of these, however, are probably better regarded as subspecies or as even less well-defined taxa. There are two clearly separate groups within the subgenus: those which are transmitted cyclically by tsetse flies (some noncyclical transmission by biting Diptera doubtless occurs in this group, but it is of minor overall importance), and those which are not. The former group consists "classically" of three species, *T. brucei*, *T. rhodesiense*, and *T. gambiense*, of which the first is a pathogen of livestock and only the last two infect man. For convenience the latter two may be regarded as separate subspecies or, more correctly, sibling species, because their geographical and hostal ranges overlap. Recent evidence from enzyme electrophoresis and characterization of variable antigenic types (see below) has been interpreted as indicating that *T. "rhodesiense"* is merely a subset of variants of *T. brucei brucei*, while *T. gambiense* can be regarded as a distinct sibling species. Probably the three really represent a single species-complex in a state of evolutionary "flux". In the second group are *T. evansi*, *T. equiperdum*, and *T. equinum*, of which *T. equiperdum* may be regarded as a sibling species of *T. evansi*; their true taxonomic relationship with each other, and with their "parent" species *T. brucei*, is at present unclear. Hereafter we shall use the term *T. brucei* s.l. (= *sensu lato*, in the broad sense) when we do not wish to differentiate between the three components of the *T. brucei* complex.

Trypanosoma brucei s.l. is pleomorphic (or polymorphic), i.e. it exists in the blood of its mammalian hosts (Table 3.4) in two morphological forms: long, slender trypomastigote individuals (about 30×1.5 µm) with a distinct free flagellum, and short, stumpy trypomastigotes (about 18×3.5 µm) with no free flagellum (Fig. 3.26). Slender forms

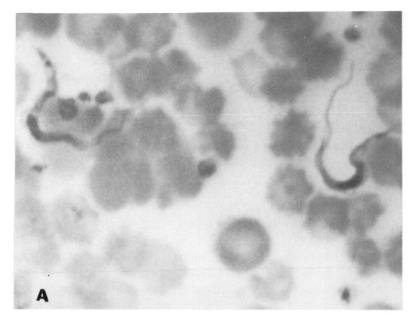

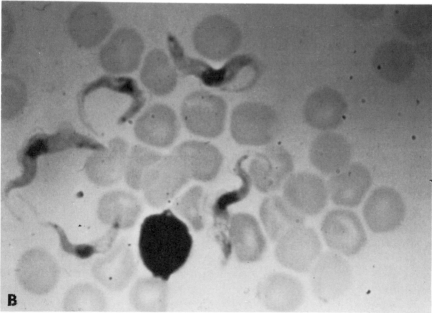

Figure 3.26 *Trypanosoma brucei gambiense*. Giemsa-stained thin blood films. The trypanosomes are pleomorphic, existing as long, slender forms (a) and as shorter, stouter forms (b).

can convert into stumpy forms, and a complete range of intermediates is seen. The relative numbers of the two forms vary widely. *Trypanosoma brucei* s.l. usually gives rise to relapsing infections in man and many domestic animals, in which peaks of parasitemia are followed by the production of antibody by the host which kills most (but not all) of the trypanosomes. The surface antigen on the survivors of this crisis differs antigenically from that on the bulk of the population before crisis; thus the survivors which are unaffected by the antibody increase in numbers again until the host has produced a second antibody, specific to this variant antigenic type (VAT), and so on. Ultimately the host almost always loses this battle, though "self-cure" of infected persons may occur rarely. It appears that anything from 100 to 1000 antigenic variants can be produced by a single antigenic "family" of trypanosomes.

There has been rapid progress in elucidating the mechanism underlying this antigenic variation since its occurrence was first established in the 1960s. The new information is extensive. In very bald outline, the blood-dwelling trypomastigotes of *T. brucei* s.l. (and probably all members of the subgenus *Trypanozoon* if not all salivarian trypanosomes) possess a glycoprotein "coat" attached to the outer surface of the plasmalemma. By shedding the molecules of this coat, and replacing them with others containing different amino acid sequences, the parasites can present an ever-changing surface antigen to the outside world (i.e. the host) and so evade destruction by the host's antibodies. Control of the synthesis of these surface coat glycoproteins (also known as variant surface glycoproteins, or VSGs) resides in nuclear genes, the sequential activation of which appears to be endogenous; it may involve gene reduplication and translocation to an "active" site near the end of the chromosome, although some VSG genes appear to be activated by a different method, not yet fully understood.

During the periods of increasing parasitemia, leading up to a crisis, trypanosomes are predominantly of the long, slender type, this being the only type which divides; as the crisis approaches, stumpy forms become more numerous, and during and immediately after it they may predominate. Sometimes a form of stumpy individual develops in which the nucleus lies in the posterior third of the body close to the kinetoplast. The significance of these "posteronuclear" (PN) forms is not known. There is fairly conclusive evidence that the stumpy forms are those which continue the developmental cycle in the tsetse fly.

The slender forms of this subgenus are unusual in having an aerobic system of oxidative metabolism which does not involve the Krebs' cycle–cytochrome chain enzymes of the mitochondrion; thus their mitochondria are non-functional. However, the stumpy forms, as well as all stages developing in the vector with the exception of the metacyclic forms, undergo normal mitochondrial terminal respiration (this supports the view that it is the stumpy forms which continue development in the vector). In the slender forms, glucose is degraded only as far as pyruvic acid, and consequently more oxygen is required for the release of a given amount of energy than in the conventional breakdown to carbon dioxide and water; thus, presumably, this type of respiration depends on the presence of a high concentration of oxygen in the environment, as obtains in blood. There is some evidence now that intracellular stages may occur in the vertebrate host, at least with certain strains of *T. brucei* s.l., but their precise significance in the life cycle is not yet fully understood.

In the tsetse fly, *T. brucei* s.l. undergoes a complicated developmental cycle in which very elongated trypomastigote forms multiply in the insect's midgut within the peritrophic membrane (a chitinous tube lining the midgut). The classical view of their subsequent development is that they then pass out of the open hind end of the peritrophic membrane, at the junction of the mid- and hindgut, and migrate forwards between the membrane and the gut wall to the proventriculus (the anterior part of the midgut), where the peritrophic membrane is secreted. They pass through the newly-secreted membrane while it is still soft, enter the lumen of the proventriculus, and then travel down the proboscis and back up the salivary duct to the salivary glands. In the glands they become attached to the epithelium by means of their expanded flagellar membranes; epimastigote forms develop, multiply, and subsequently metamorphose into metacyclic trypomastigote forms which become detached and are injected into the next vertebrate on which the fly feeds; if this belongs to a susceptible species of mammal, they change into the slender blood forms and begin an infection. Only a small proportion of tsetse flies which ingest *T. brucei* s.l. develop salivary gland infections, which is perhaps not surprising in view of the complicated migrations supposedly undergone by the trypanosomes in the fly. Recent evidence, however, has cast considerable doubt on the truth of the details of this odyssey. Intracellular stages in the tsetse's midgut wall have been reported, followed by a more direct migration to the salivary glands via the hemocoel, as occurs with *T. rangeli*.

Some strains or isolates of *T. brucei* s.l. (the classical "*T. brucei*" itself) are unable to infect man, but infect wild game and cause disease in domestic animals. The stability of this character is in doubt. As we have already mentioned, recent biochemical evidence indicates that one of the subtaxa of *T. brucei* which infects man is a sufficiently coherent and stable entity to justify its being regarded as a separate sibling species – *T. brucei gambiense*. However, the other subtaxon, *T. brucei rhodesiense*, is probably only a relatively unstable subset of enzymic or antigenic variants (or both) of *T. brucei*, perhaps best distinguished (when necessary) as the rhodesiense clinical variant, or nosodeme. The use of the word "deme" (from Greek *demos*, "a population") as a descriptive suffix was introduced into trypanosome taxonomy by Hoare. It has proved useful, with an increasing number of prefixes, to delineate various kinds of subspecific populations within the *T. brucei* complex. Nosodeme, for clinically-defined populations, is one of the older; others include zymodeme, for populations defined by their isoenzyme constitution; serodeme, for populations defined by their VAT repertoire; peptideme, for populations defined by polypeptide constitution; and schizodeme, for populations defined by restriction endonuclease analysis of their DNA (the last term is undesirable, since it refers, unlike the others, not to the distinguishing characteristic but to the technique used to determine it). Thus the rhodesiense population can be regarded as either a nosodeme or a group of zymodemes of *T. brucei*.

Whatever the precise terminology one adopts, African human trypanosomiasis occurs as one of two, admittedly rather ill-defined and overlapping, clinical and geographical variants throughout tropical Africa. The gambiense form occurs in West and Central Africa, and the rhodesiense form in the eastern half of that continent. Around the northern and eastern shores of Lake Victoria, the two groups meet and overlap. The diseases produced in man by the two populations are basically similar, differing in that gambiense infections are much more chronic, lasting a matter of years in untreated patients, while those due to rhodesiense are measured in months. Both forms are fatal unless treated (with possible rare exceptions). Initially the trypanosomes live and multiply in the blood and tissue fluid, producing a febrile condition which may be quite mild. After a few months or a year or so, the parasites invade the central nervous system and multiply in the cerebrospinal fluid; here, they ultimately cause brain damage which leads to coma, from which the disease gets the name "sleeping sickness", and death.

The gambiense parasites are usually transmitted to man by *Glossina palpalis* and related species, flies which live mainly along the banks of rivers; the rhodesiense parasites are usually transmitted by flies of the *Glossina morsitans* group, which inhabit vast areas of the drier plains in East Africa (Fig. 3.27). This results in differences in the epidemiology of the two diseases.

Trypanosoma evansi, *T. equiperdum*, and *T. equinum* (see Tables 3.2 & 3.4) are thought to have evolved relatively recently from *T. brucei* s.l. by the carriage of the latter in infected animals (possibly camels) northwards in Africa beyond the range of *Glossina*. Here, the parasites became adapted to noncyclical transmission by biting Diptera (cf. *T.*

Figure 3.27 *Trypanosoma (b.) gambiense*, the West African form of African sleeping sickness, is usually transmitted by *Glossina palpalis* and related species which live in moist areas, mainly along river banks (a). The rhodesiense nosodeme is usually transmitted by flies of the *G. morsitans* group which inhabit drier areas (b).

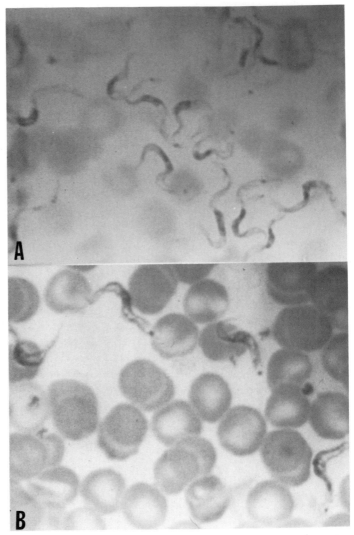

Figure 3.28 The two organisms *Trypanosoma evansi* (a) and *T. equiperdum* (b) are monomorphic trypanosomes of the *brucei* group. They are morphologically very similar and resemble the long, thin forms of *T. brucei*.

vivax in Mauritius), and so survived and underwent speciation on their own.

Strains of *T. brucei* s.l. which are transmitted by biting flies such as Tabanidae, or by hypodermic syringes in laboratories, no longer need the ability to infect tsetse flies. Selection pressure operates against the retention of pleomorphism in such strains, as the stumpy forms are not only unnecessary but, since they do not divide, actually reduce the strain's reproductive rate and thus diminish its chances of being transmitted to another vertebrate host and, hence, surviving. Noncyclically transmitted strains in laboratories, and by inference in the field, quite rapidly (within a few years at most) become monomorphic, all individuals being more or less of the slender type (Fig. 3.28). Such strains also lose the ability to infect tsetse flies, presumably because they have lost the ability to synthesize mitochondrial enzymes. This is presumably how *T. evansi* evolved, since it is usually monomorphic (being morphologically indistinguishable from a strain of *T. brucei* s.l. which has been maintained by syringe-passage in a laboratory for some years) and unable to infect tsetse flies: in this instance the biting Diptera of North Africa played the role of the laboratory syringe.

This hypothetical but plausible evolutionary story can be carried further. The kinetoplast (a mass of DNA within the mitochondrion) is presumably concerned, at least in part, with the synthesis of the mitochondrial enzymes. Thus, in the slender forms of *T. brucei* it is redundant, at least as far as this function is concerned; and it is common to find a small proportion of the slender forms of this species (about 1%) lacking a conventionally organized kinetoplast. In strains transmitted cyclically by tsetse flies, such dyskinetoplastic individuals would be "filtered out" by their inability to synthesize the necessary respiratory enzymes to enable them to develop in the insect. However, in *T. evansi* this particular selection pressure would be lacking, and indeed strains of this species are often found with quite high proportions of dyskinetoplastic individuals. Sometimes, particularly in

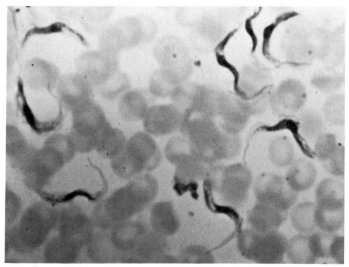

Figure 3.29 *Trypanosoma equinum* resembles the long, slender forms of *T. brucei*. It differs from the other members of the group, however, in lacking a normal kinetoplast.

85

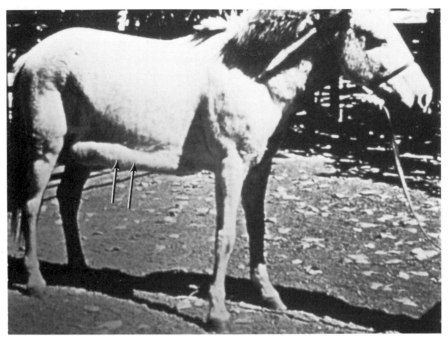

Figure 3.30 In dourine of the horse, the trypanosomes develop in large numbers in the subcutaneous and submucosal tissues causing edema (arrows). The trypanosomes pass from one horse to another with edematous fluid released from abrasions of the genitalia during sexual intercourse.

South America, strains are found in which all the organisms lack a kinetoplast (Fig. 3.29). These are generally referred to as a distinct species, *T. equinum*.

Trypanosoma equiperdum is morphologically identical with *T. evansi* but has become specialized in another direction, being the only known member of the genus which is not dependent upon an invertebrate vector to transmit it from vertebrate to vertebrate. It is a parasite of horses and donkeys, and is more numerous in edematous patches of skin than in the blood (Fig. 3.30). When such patches occur on the genitalia, they often become eroded during coitus and it is in this way that the trypanosomes are transmitted.

DIAGNOSIS OF TRYPANOSOME INFECTIONS

The simplest method of diagnosing trypanosome infections of vertebrates is by making and examining thin films of peripheral blood stained with Giemsa's stain, at a magnification of × 500–× 1000. Alternatively, a drop of blood can be mounted on a slide beneath a coverslip and examined immediately for motile trypanosomes at a magnification of about × 400. If the parasites are scanty, they may not be found by these methods. Some concentration of parasites in the blood of mammals can be achieved by making stained thick blood films. Concentration is also achieved by centrifugation of blood in a heparinized capillary tube. The trypanosomes concentrate just above the red cells. The tube is scored and broken and the serum at that point examined. The most sensitive method of diagnosis, for most species of *Trypanosoma*, is to collect some blood aseptically and inoculate it to blood–agar cultures; trypanosomes, if present, will multiply as (usually) epimastigotes. However, the Salivaria are difficult to cultivate in this way and special media must be used.

Apart from these methods, human infections with *T. cruzi* may be detected by feeding reduviid bugs, which have been bred in the laboratory, on a suspect and then seeing whether the bugs become infected. This method is known as **xenodiagnosis**. *Trypanosoma brucei gambiense*, which may be very sparse in the peripheral blood of infected persons, is often more readily found in the "juice" obtained by puncturing lymph glands. In the later stages of infection with both *T. b. gambiense* and "rhodesiense", organisms may be found more readily in cerebrospinal fluid obtained by lumbar puncture than in blood. Inoculation of blood (or other material) to susceptible laboratory animals such as rats or mice is also used. *T. b. gambiense* is usually only slightly, if at all, infective to mature animals; young (weanling) rats are more susceptible.

Immunodiagnostic methods are becoming increasingly used, especially in suspected African trypanosomiasis. The techniques used include indirect immunofluorescence, enzyme-linked immunosorbence assay (ELISA), and agglutination. More details are given in the paper by De Raadt and Seed, cited in the "Further Reading" list at the end of this chapter.

TREATMENT OF LEISHMANIAL AND
TRYPANOSOMAL INFECTIONS

Treatment of leishmaniasis caused by members of the *L. donovani* complex is with compounds of antimony, and is often, but not always, effective. It is sometimes followed by a nodular eruption of the skin which may be allergic in origin, called post-kala azar dermal leishmanoid; amastigotes are found in the nodules. Treatment of leishmaniasis caused by members of the *L. braziliensis* group is difficult.

Treatment of leishmaniasis caused by *L. tropica* need be undertaken only if normal healing is delayed.

No really effective drug is available for curing or treating Chagas disease, though some are being used clinically and others are under test.

In the early stage of African sleeping sickness, before invasion of the central nervous system, the disease is usually curable (the most commonly used drug is suramin). In the late stage, treatment is less successful, and is based on rather toxic organic arsenical compounds like melarsoprol (= melarsen oxide/BAL, or mel B; "BAL" or "B" refer to "British anti-lewisite, or 2,3-dimercaptopropanol). Better drugs are needed. A few of the drugs used in treatment of leishmanial and trypanosomal infections are listed in Table 3.5.

The drugs used to treat or control trypanosomiasis of domestic livestock include quinapyramine ("Antrycide"), also commonly used as a chemoprophylactic; diminazene ("Berenil"); various phenanthridinium derivatives – homidium ("Ethidium"), pyrithidium ("Prothidium"), isometamidium ("Samorin"); and suramin (also used to treat human trypanosomiasis).

The development of a vaccine against African human trypanosomiasis, currently the "holy grail" of many immunoparasitologists, is bedevilled by the occurrence of antigenic variation (discussed above). Its attainment on anything other than a limited, experimental level seems unlikely at the time of writing (1986).

Table 3.5 Some drugs used in treatment of leishmanial and trypanosomal infections of man.

Leishmania spp.	Pentostam (sodium stibogluconate) or Glucantime (meglumine antimonate); both contain pentavalent antimony: pentamidine isethionate or amphotericin B for refractory or relapsing infections
Trypanosoma cruzi	Nifurtimox (a nitrofuran) or benznidazole (a nitroimidazole) are the only available drugs; neither is reliable and both are very toxic.
Trypanosoma brucei s.l. chemoprophylaxis	pentamidine isethionate*
early-stage infections (normal cerebrospinal fluid)	pentamidine isethionate* or suramin sodium
late-stage infections (abnormal cerebrospinal fluid)	melarsoprol; nitrofurazone (a nitrofuran) for refractory or relapsing infections

* Not recommended for the "rhodesiense" nosodeme.

FURTHER READING

Bernards, A., T. Delange, P. A. M. Michels, A. Y. D. Lin, M. J. Huisman & P. Borst 1984. Two modes of activation of a single surface antigen of *Trypanosoma brucei*. *Cell* **36**, 163–70.

De Raadt, P. & J. R. Seed 1977. Trypanosomes causing disease in man in Africa. In *Parasitic Protozoa*, Vol. 1, J. P. Kreier (ed.), 176–237. New York: Academic Press.

Elliott, K., M. O'Connor & G. Wolstenholme (eds) 1974. *Trypanosomiasis and leishmaniasis with special reference to Chagas' disease*. Ciba Foundation Symposium 20. Amsterdam: Elsevier.

Hoare, C. A. 1967. Evolutionary trends in mammalian trypanosomes. *Advances in Parasitology* **5**, 47–91.

Hoare, C. A. 1972. *The trypanosomes of mammals; a zoological monograph*. Oxford: Blackwell Scientific Publications.

Hudson, L. (ed.) 1985. *The biology of trypanosomes*. Current topics in microbiology and immunology, no. 117. Berlin: Springer-Verlag.

Koberle, F. 1968. Chagas' disease and Chagas' syndromes: the pathology of American trypanosomiasis. *Advances in Parasitology* **6**, 63–116.

Kolata, G. 1984. Scrutinizing sleeping sickness. *Science* **226**, 956–9.

Lumsden, W. H. R. & D. A. Evans (eds) 1976 & 1979. *Biology of the Kinetoplastida*, 2 vols. London: Academic Press.

Molyneux, D. H. & R. W. Ashford 1983. *The biology of* Trypanosoma *and* Leishmania, *parasites of man and domestic animals*. London: Taylor & Francis.

Morrison, W. I., M. Murray & W. I. M. McIntyre 1981. Bovine trypanosomiasis. In *Diseases of cattle in the tropics*, M. Ristic & I. McIntyre (eds), 469–97. The Hague: Martinus Nijhoff.

Murray, M., W. I. Morrison & D. D. Whitelaw 1982. Host susceptibility to African trypanosomiasis: trypanotolerance. *Advances in Parasitology* **21**, 2–68.

Newton, B. A. (ed.) 1985. Trypanosomiasis. *British Medical Bulletin* **38**, no. 2., 103–99.

Society of Protozoologists 1984. Symposium – antigenic variation in trypanosomes. *Journal of Protozoology* **31**, 41–74.

Tait, A. 1980. Evidence for diploidy and mating in trypanosomes. *Nature* **287**, 536–8.

Tizard, I. (ed.) 1985. *Immunology and pathogenesis of trypanosomiasis*. Boca Raton, Florida: CRC Press.

Turner, M. J. 1982. Biochemistry of variant surface glycoproteins of salivarian trypanosomes. *Advances in Parasitology* **21**, 70–153.

Turner, M. J. 1983. Antics of the elusive trypanosome. *Nature* **303**, 202–3.

WHO Expert Committee Report 1984. *The leishmaniases*. (World Health Organization Technical Report Series, no. 701). Geneva: World Health Organization.

Zuckerman, A. & R. Lainson 1977. *Leishmania*. In *Parasitic Protozoa*, Vol. 1, J. P. Kreier (ed.), 58–133. New York: Academic Press.

CHAPTER FOUR

Flagellates of the alimentary and urinogenital tracts

Seven of the eight orders of the class Zoomastigophorea (subphylum Mastigophora, phylum Sarcomastigophora) are mostly or entirely parasitic. One parasitic order, the Kinetoplastida, forms the subject of the preceding chapter, the rest will be considered here, together with the subphylum Opalinata.

CLASS ZOOMASTIGOPHOREA

There are many symbiotic genera in the Zoomastigophorea; very few are of economic importance, and probably only four genera (*Histomonas*, *Hexamita*, *Giardia*, *Trichomonas*) contain pathogenic species. Some, in the order Hypermastigida, are not only useful but essential to their hosts (termites and certain Orthoptera or wood-roaches). Most species are transmitted directly by the voiding of parasites (often encysted) in the feces and their ingestion by a new host; exceptions to this simple system will be mentioned below. Sexual reproduction is known in only a few genera of the orders Oxymonadida and Hypermastigida.

ORDER RETORTAMONADIDA

Several genera of small symbiotic flagellates with 2–4 flagella are included in this order. They have a cytostome, are not bilaterally symmetrical and lack an axostyle and an undulating membrane. None is of economic importance, nor, as far as is known, pathogenic. Their hosts may be vertebrate or invertebrate. Two species inhabit the intestine of man, *Retortamonas intestinalis* (Fig. 4.1) and *Chilomastix mesnili* (Fig. 4.2), (Table 4.1)

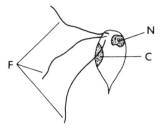

Figure 4.1 *Retortamonas* sp. Retortamonads are 5–20 μm in diameter and have 2–4 flagella (F), one of which is recurved and is associated with the cytostomal (C) region. A nucleus (N) is present in the anterior region of the parasite's body.

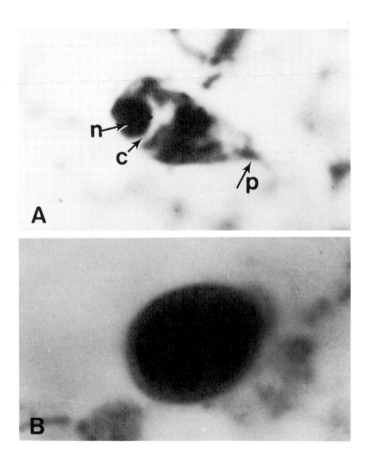

Figure 4.2 *Chilomastix mesnili* (a). Photomicrograph of a vegetative form in a stained, fixed fecal smear. The pointed posterior end (P), nucleus (N) and cytostomal groove (C) are visible in this picture but the flagella are not. The cyst (b) is lemon-shaped.

Table 4.1 Flagellates inhabiting the human intestine

Species	Trophozoite			Cyst		
	size (μm)	no. of flagella	other characters	size (μm)	no. of nuclei	other characters
Chilomastix mesnili	6–24 × 3–10	3+1*	cytostome	7–10	1	lemon-shaped; cytostome fibril
Enteromonas hominis†	4–10 × 3–6	3+1	no cytostome	6–8 × 4–5	4	none
Retortamonas intestinalis	4–9 × 3–4	1+1	cytostome	4.5–6	1	as *C. mesnili*
Giardia lamblia‡	10–20 × 5–10	4 pairs	attachment disc; 2 nuclei; median bodies	8–14 × 6–10	4	fibrils and axonemes
Trichomonas hominis	8–20 × 3–14	5+1	undulating membrane; axostyle; cytostome			no cyst

* This notation implies that there are three free anterior flagella and one recurrent one.
† The taxonomic position of this organism within the Zoomastigophorea is doubtful.
‡ Sometimes pathogenic.

ORDER OXYMONADIDA

These flagellates, which have four or more flagella and one to many nuclei and axostyles, are all symbiotic in the intestine of insects – roaches (Orthoptera, Blattidae) or termites (Isoptera).

ORDER DIPLOMONADIDA

This order includes flagellates with a bilaterally symmetrical body, two nuclei and up to eight flagella. Almost all are symbiotic in vertebrates or invertebrates. The two main genera, *Hexamita* and *Giardia*, are most easily distinguished by the fact that the former is narrower, lacks an anterior attachment disc, and has two pairs of flagella emerging from the body at the extreme front end; it does not encyst. *Giardia* is described below. *Hexamita spp.* inhabit the intestines of amphibia, fish, birds, and orthopterous insects. *Hexamita salmonis* (of fish), *H. meleagridis* (of turkeys), and *H. columbae* (of pigeons) are pathogenic. Another genus, *Spironucleus*, infects rodents.

GENUS *GIARDIA* KUNSTLER, 1882

Species of this genus have been described from the small intestines of man, dogs, cats, cattle, various rodents, rabbits, and other mammals; a few species are recorded from amphibia and reptiles. How many of these are really valid species is unknown, as is the pathogenicity of most of them. However, *G. canis* of the dog probably sometimes behaves as a pathogen, and *G. lamblia* of man certainly does so at times. The general morphology of all species is familiar, and all are transmitted by means of cysts passed out in the feces.

Giardia lamblia

There is much confusion about this organism's correct name; it is also known as *G. intestinalis*, *G. duodenalis* and even as *Lamblia intestinalis*. It has the distinction of being probably the first symbiotic protozoon to be seen – by Antony van Leeuwenhoeck in 1681, in his own feces.

Giardia lamblia inhabits the lumen of the duodenum and upper ileum of man, monkeys and pigs in all parts of the world. In man it is common (about 10%), especially in children. It can cause a disease known as giardiasis or lambliasis. Beavers (*Castor canadensis*) may serve as a reservoir host for human infection in North America, and infections have occurred among persons camping in relatively isolated areas.

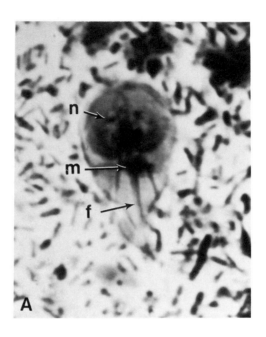

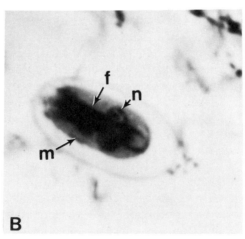

Figure 4.3 *Giardia lamblia* trophozoite (a) and cyst (b). The trophozoite in this photomicrograph of a fecal smear is in plan view. The attachment disc (sucker) is the central structure around the nuclei (n). The median body (m) is just below the sucker, and the flagella (f), where they pass within the body, are also visible. In the cyst (b) the various organelles form a jumble in the cytoplasm within the thick cyst wall.

Morphology and life cycle The trophozoite is shaped like a pear bisected longitudinally (Fig. 4.3). The flat (ventral) surface of the thick, broad (anterior) end forms an attachment disc (conventionally called a "sucker"), with a thickened anterior rim. By means of the disc, perhaps aided by a lectin activated by the host's proteases, the organism attaches to the intestinal mucosa. It has been proposed that the activity of the ventral flagella (see below) creates negative hydrostatic pressure beneath the disc, thus keeping the organism attached. Above the disc are two nuclei, and between the latter lie eight basal bodies giving rise to eight flagella. Only two of the flagella emerge directly from the body; two cross over and follow the front edge of the disc emerging laterally, and the remaining four run backwards in the body for some way – two (the ventral pair) emerging at the hind end of the disc, and two at the extreme hind end of the body. By means of these flagella the trophozoite can swim actively. In the posterior half of the body lie one or (usually) two curved "median bodies" of unknown function (often incorrectly called parabasal bodies). The trophozoites are usually 10–20 μm long, 5–10 μm broad, and 2–4 μm thick. Reproduction is by binary fission; no sexual stage is known.

The cysts of *G. lamblia* are oval, about 8–14 μm long and 6–10 μm wide, and contain (when mature) four nuclei grouped at one end. The remains of the median body, flagella, and the anterior rim of the attachment disc form a rather confused collection of fibrils within the cyst (see Fig. 4.3), which makes it easy to recognize in fecal preparations. When a cyst is swallowed by a susceptible host, it presumably hatches in the duodenum. The quadrinucleate organism which emerges then divides into two binucleate trophozoites.

Pathogenesis *Giardia lamblia* does not invade the tissues; heavy infections may produce acute but not bloody diarrhea, especially in children, and epigastric pain. The parasites are thought to swim up the bile duct and into the gall bladder where they may produce jaundice, nausea, and vomiting. In heavy infections, the attachment of a large number of organisms to the intestinal epithelium (Fig. 4.4) may interfere mechanically with the absorption of fat (and fat-soluble vitamins).

Diagnosis Diagnosis is confirmed by finding cysts, which may be very numerous, and, in cases with diarrhea, trophozoites in fecal specimens.

Treatment and prevention Giardiasis is said to be readily cured by 5-nitroimidazoles such as metronidazole or tinidazole, though some infections may be surprisingly resistant. Prevention is solely a matter of

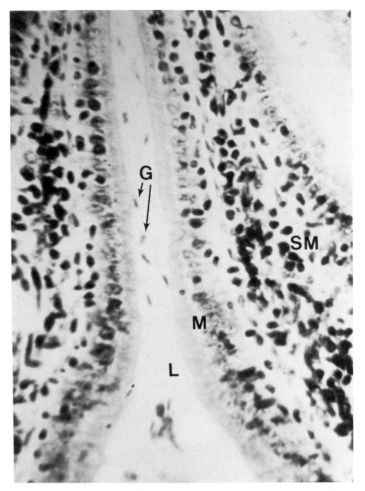

Figure 4.4 Photomicrograph of *Giardia lamblia* in a section of intestine. The giardia (G) do not penetrate the mucosa but remain in the lumen (L). The mucosal lining (M) and submucosal tissue (SM) appear normal in this section.

hygiene, particularly with regard to food and water, to prevent the ingestion of cysts.

ORDER TRICHOMONADIDA

This order includes forms with, typically, 3–5 free anterior flagella and one recurrent flagellum which may be attached to the body to form an undulating membrane. Members of one genus (*Histomonas*) have only one flagellum and members of another have no flagellum (*Dientamoeba*).

Almost all are parasitic in vertebrates or invertebrates, including termites, in which some species are truly mutualistic and help to digest the wood on which the termite feeds but is itself unable to digest. No member of this order is known to produce cysts, transmission being usually by the ingestion of resistant trophozoites passed out in the feces of infected hosts. The only economically important parasites in this order belong to the genera *Trichomonas* and *Histomonas*.

GENUS *TRICHOMONAS* DONNÉ, 1837

Species of this genus occur in the intestines of mammals including man (Table 4.1), birds, reptiles, amphibia, molluscs (slugs), and termites; in the mouth of man and monkeys (*T. tenax*); and in the urinogenital tract of man and cattle (*T. vaginalis* and *T. foetus*, respectively). No cyst is produced, and no sexual process is known. Reproduction is solely by binary fission.

Some authors divide the genus *Trichomonas* into three subgenera, or even separate genera, on the basis of the number of anterior flagella. Species with three (including *T. foetus*) are called *Tritrichomonas*, those with four (including *T. vaginalis*, *T. tenax* and *T. gallinae*) are called *Trichomonas*, and those with five (including *T. hominis*) are called *Pentatrichomonas*. The following three species are of medical and economic importance.

Trichomonas vaginalis

This organism seems to be naturally exclusive to our own species (though hamsters can be infected experimentally). It lives in the female vagina and the male urethra or prostrate, and is common throughout the world, particularly in women. Infection rates of up to forty percent of unselected women have been recorded and the prevalence among those with vaginal upsets is higher – up to seventy percent. *Trichomonas vaginalis* may cause an annoying but not serious disease in women called trichomonas vaginitis.

Morphology and life cycle   *Trichomonas vaginalis* is ovoid, narrower at the hind end, ranging in size from 10 to 30 μm long (usually averaging 14–17 μm) and 5–15 μm broad (Fig. 4.5). It has four free anterior flagella and one recurved posterior one attached to a thin, fin-like extension of the body to form an undulating membrane. The recurved flagellum and undulating membrane end about half way along the body. All five flagella arise from basal bodies grouped at the anterior end, just in front of the single nucleus. With the aid of these

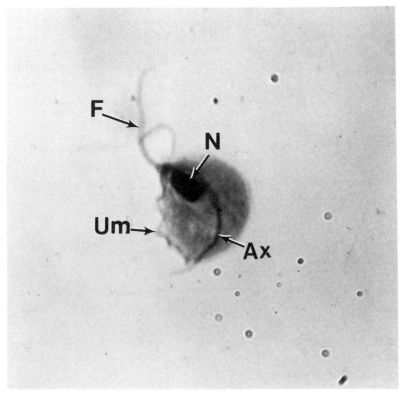

Figure 4.5 *Trichomonas vaginalis*, photomicrograph of a stained organism. The nucleus (N), flagella (F), undulating membrane (Um) and axostyle (Ax) are all visible in this preparation.

flagella, the organism can swim actively. It also has a prominent skeletal axostyle of longitudinally arranged parallel microtubules, which may protrude from the hind end of the body; a cytostomal groove; a supporting costa at the base of the undulating membrane; and a true parabasal body beside the nucleus. When seen alive the parasite is easily identified by the characteristic and beautiful progression of waves backwards along the undulating membrane.

Transmission is usually by sexual contact, the infected male serving as a reservoir, but may also occur by contamination of the vaginal orifice with infected material.

Pathogenesis   Most commonly the parasite is nonpathogenic; this is almost always so in males, though mild inflammation of the urethra may occasionally result. In women the organism is rather more commonly responsible for mild vaginal inflammation associated with a copious, foul-smelling discharge. It has been suggested that the parasite's pathogenicity is associated with endocrinal or other changes resulting in variation in the normal bacterial flora of the vagina, leading to a reduction in the acidity of its contents from the usual pH 4–4.5 to pH 5.5. The organism cannot survive at neutrality (pH 7). *Trichomonas vaginalis* does not invade the tissues as far as is known.

Diagnosis   Diagnosis is confirmed by isolating organisms from the vaginal discharge. This is best done by cultivation *in vitro*. Microscopic examination is less reliable, though fresh preparations in which the organisms are motile are better than dried smears stained with Giemsa's (or other) stains.

Treatment   Treatment is possible, though radical cure is sometimes not easily attained. Drugs of the 5-nitroimidazole family such as metronidazole, given orally, are generally effective.

Trichomonas foetus

This species infects cattle and possibly other farm animals. *Trichomonas suis* of pigs is similar to *T. foetus* and it is possible that *T. foetus* is a strain of *T. suis* adapted to cattle. *Trichomonas foetus* lives in the vagina and uterus of the cow and in the prepucial sheath of the bull. It is widely distributed throughout the world and is probably fairly common, but details of its distribution and prevalence are not known. Replacement of natural breeding by artificial insemination with semen from uninfected bulls has reduced the incidence of infection.

Morphologically, *T. foetus* is basically similar to *T. vaginalis* but it has only three anterior flagella and the recurrent flagellum extends free for some distance beyond the hind end of the body. The trophozoite is transmitted during coitus or by artificial insemination with contaminated semen. If it is present in a pregnant uterus the organism invades the fetus and an early abortion often results (usually between 1 and 16 weeks after fertilization). Diagnosis is confirmed in the same ways as for human *T. vaginalis* infection.

In cows the infection is usually self-limiting, all trophozoites being expelled with the aborted fetus and placenta. This is fortunate since no reliable treatment is known. If untreated, bulls remain infected for life. Treatment of bulls by oral administration of metronidazole has been reported to cure infection. It may be used in an attempt to save valuable bulls, but less-valuable bulls are best destroyed.

Trichomonas gallinae

This species infects birds, primarily the domestic pigeon *Columba livia* but also other Columbiformes, turkeys, and chickens. The trophozoites inhabit the mouth, pharynx, esophagus, and crop. *Trichomonas gallinae* is common in pigeons but rare in chickens. The organism resembles *T. vaginalis* but is smaller, 6–19 × 1–9 μm. There are four anterior flagella and the recurrent flagellum ends about three-quarters of the way along the body.

Trichomonas gallinae produces a severe disease in some young pigeons, which may be fatal. The disease is variously known as canker, frounce, roup, or, more scientifically, avian trichomoniasis. Almost all adult birds of susceptible species are infected but show no sign of disease. The organisms produce lesions in the anterior part of the digestive tract, and may spread to the sinuses of the head and to viscera such as the heart, lungs, and liver. The lesions appear as small yellowish spots on the mucosa, and develop into large caseous nodules or masses. Young pigeons become infected from their parents during feeding (pigeons regurgitate food from their crop into the nestling's mouth).

Diagnosis is confirmed by finding trophozoites either in smears or in cultures made from lesions. The disease can be treated with 2-amino-5-nitrothiazole, one of the drugs used in treating histomoniasis. The only way to keep young birds free from infection is to treat the adults before the eggs hatch.

GENUS *DIENTAMOEBA*

Until recently *Dientamoeba* was considered to be an aberrant ameba. It, like trichomonads, has two nuclei in the vegetative form and produces no cysts. The only member of the genus, *D. fragilis* (Fig. 4.6), inhabits the colon of man where it apparently causes no harm.

The transmission of *D. fragilis* is puzzling since the organism produces no cysts and the trophozoite is unlikely to survive a journey through the acid stomach contents, even if it could live long enough outside one host to have a good chance of reaching a second. It is possible (though there is no direct evidence supporting this), that it travels from host to host as a passenger inside the eggs of a nematode as other protozoa are known to do (e.g. *Histomonas*); a possible candidate vector is *Enterobius*.

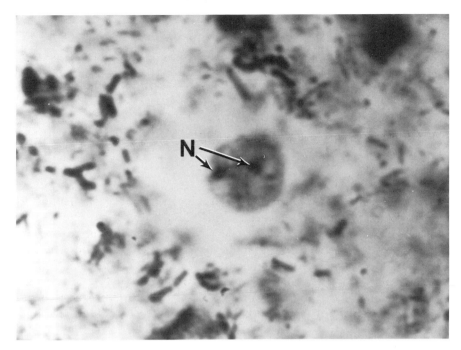

Figure 4.6 Photomicrograph of *Dientamoeba fragilis*. The two nuclei (N) characteristic of this organism are readily seen in this preparation in a stained fecal smear.

GENUS *HISTOMONAS* TYZZER, 1920

Only one species is known in this genus – *H. meleagridis*. It inhabits the intestinal ceca (both lumen and mucosa) and the liver of gallinaceous birds, including chickens and turkeys, and has been reported from all parts of the world. In chickens it is seldom pathogenic and extremely common. In turkeys it is one of the more important pathogens, producing a disease called histomoniasis or "blackhead" (because the head of an infected bird may become darkened). Electron microscopic study of the structure of *H. meleagridis*, which revealed a relationship to the trichomonads, is the basis for the transfer of the genus from the Sarcodina to the Mastigophora.

Histomonas meleagridis

Morphology and life cycle This is a flagellate organism which may also have a non-flagellate phase. The latter resembles an ameba and is found only in the host's tissues (Fig. 4.7); it is usually oval, in the size range 8–21 × 6–15 µm. The flagellate phase occurs at times in the

101

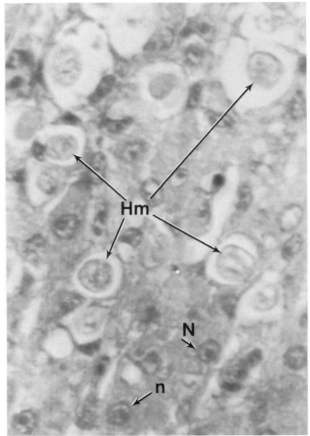

Figure 4.7 Photomicrograph of *Histomonas meleagridis* in a liver section. Ameboid forms (Hm) are present in vacuoles in the liver tissue. Nuclei of liver cells (N, n) can also be seen.

lumen of the ceca (and in cultures *in vitro*); it, too, is ameboid, from 5 to 30 μm in diameter, and possesses usually one short flagellum (sometimes as many as four), arising from a basal body (or bodies) close to the single, eccentric nucleus. The parasites divide by binary fission, no sexual process being known.

The trophozoites do not survive well outside the body and no cyst is produced. Transmission occurs by means of trophozoites carried inside the eggs of the parasitic nematode *Heterakis gallinae*. By no means all of the worms from an infected host carry the protozoon, possibly only about 0.1%. However, infected eggs can survive in soil for several years.

Epidemiology   As stated above, *Histomonas* is very common in chickens, to which it is rarely pathogenic. *Heterakis* is also common in these birds. It is probable that turkeys become infected by ingesting infected eggs deposited by chickens. Wild gallinaceous birds (e.g. pheasant, quail, grouse) may sometimes also serve as reservoirs of infection.

Pathogenesis   When the non-flagellate stages invade and multiply in the cecal mucosa, they produce much tissue damage resulting in ulcers which may become widespread. The infected ceca become inflamed, enlarged and filled with a yellowish hard exudate; the lesions may perforate and cause peritonitis. A yellow diarrhea develops. Trophozoites are often carried by the bloodstream to the turkey's liver (and, less commonly, to other organs), where they produce circular, yellowish-green necrotic abscesses up to 1 cm or more in diameter. Trophozoites are readily found in sections of both cecal and hepatic lesions.

If untreated, death may occur rapidly in 50–100% of young turkeys, less in older birds. Birds which do recover are immune to reinfection.

Diagnosis   Apart from the clinical picture, histomoniasis can be diagnosed from the appearance of the hepatic lesions at necropsy, and confirmed by demonstrating the organisms in sections of, or scrapings from, cecal or hepatic lesions.

Treatment and prevention   Several drugs, usually given mixed with the food, may be used to suppress the infection; as radical cure is seldom achieved, treatment may be necessary throughout the bird's life. Reasonably effective drugs include 4-nitrophenylarsonic acid, 2-amino-5-nitrothiazole and derivatives, and, as a prophylactic only, furazolidone. To reduce the likelihood of their ingesting infected eggs of *Heterakis*, turkeys should not be reared together with chickens, nor on land previously used by chickens.

ORDER HYPERMASTIGIDA

Two suborders are recognized in the order Hypermastigida: the Lophomonadida, with only one anterior cluster of flagella, and the Trichonymphina, with two (rarely four) clusters of flagella. These organisms live in the gut of insects, either termites (Isoptera) or roaches (Orthoptera, family Blattidae). They are uninucleate but may have a very complex structure. Very large numbers may be present – it has been estimated that up to one-third or one-half of the insect's weight may be accounted for by its flagellates. Most of these flagellates can digest cellulose, and without them the wood-eating insects (termites and wood-roaches) would be unable to survive since they themselves cannot digest the wood on which they feed. This is a truly mutualistic relationship since the termite (or roach) obtains the food, the flagellate digests it, and both insect and protozoon live on it; neither can survive without the other. This also applies to some of the oxymonadid, diplomonadid, and trichomonadid flagellates inhabiting the gut of termites.

In termites, one of the most important genera is *Trichonympha* (Fig. 4.8); without it the insect cannot survive. Several other genera live in the termite gut, but their removal does not interfere significantly with the termites' metabolism.

The hypermastigid flagellates of termites do not encyst. Every time the growing insect moults, its infection is lost. In order to survive, it must somehow become reinfected from its neighbors (probably by ingesting flagellates in feces taken directly from the anus of another termite). In the wood-eating roaches the infection, once established, is permanent; at least some of the protozoa of these insects do form cysts by means of which other newly hatched roaches are doubtless infected. Sexual reproduction occurs in many species of these flagellates; in those of at least one genus of the wood-roach (*Cryptocerus*) it has been shown that the onset of sexuality is related to the secretion by the host insect of moulting hormone.

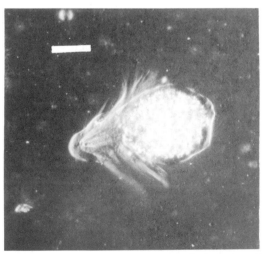

Figure 4.8 *Trichonympha chattoni*, photomicrograph of a living specimen. This flagellate occurs in the guts of termites in Florida (the bar indicates 10 μm; the photomicrograph was provided by D. Bermudes, Department of Biology, Boston University).

SUBPHYLUM OPALINATA

These organisms were the subject of much taxonomic confusion. They are large, oval, flattened protozoa, covered with cilia, and possessing two or many similar nuclei. Almost all are symbionts of the large intestine of frogs and toads (Amphibia, Salientia), a few having been described from other amphibia, reptiles, and fish. They were for many years regarded as primitive ciliates (Protociliata), but recognition of the fundamental similarity between cilia and flagella has led many people to regard the Opalinata as closer to the Mastigophora, or at least intermediate between them and the ciliates. Opalinata differ from ciliates as follows:

(a) their nuclei are not differentiated into macro- and micronuclei;
(b) sexual reproduction involves the fusion of two dissimilar flagellated individuals (anisogametes), followed by encystment, and not conjugation as in ciliates;
(c) asexual division is usually symmetrogenic as in flagellates; however, homothetogenic fission also occurs.

FURTHER READING

Holberton, D. V. 1974. Attachment of *Giardia* – a hydrodynamic model based on flagellar activity. *Journal of Experimental Biology* **60**, 207–21.

Honigberg, B. M. 1970. Protozoa associated with termites and their role in digestion. In *Termites*, Vol. 2, K. Krishna & F. M. Weesner (eds), 1–36. New York: Academic Press.

Honigberg, B. M. 1978. Trichomonads of veterinary importance. In *Parasitic protozoa*, Vol., J. P. Kreier (ed.), 163–273. New York: Academic Press.

Honigberg, B. M. 1978. Trichomonads of importance in human medicine. In *Parasitic protozoa*, Vol. 2, J. P. Kreier (ed.), 275–454. New York: Academic Press.

Jakubonski, W. & J. C. Hoff (eds) 1979. *Waterborne transmission of giardiasis*. Report number EPA-600/9-79-001. Springfield, Virginia: National Technical Information Service, US Department of Commerce.

Jírovec, O. & M. Petrů 1968. *Trichomonas vaginalis* and trichomoniasis, *Advances in Parasitology* **6**, 117–88.

Kreier, J. P. & I. E. Selman 1981. Actinobacillosis, actinomycosis, nocardiosis, eperythrozoonosis, haemobartonellosis, and trichomoniasis. In *Diseases of cattle in the tropics*, M. Ristic & I. McIntyre (eds), 567–83. The Hague: Martinus Nijhoff.

Kulda, J. & E. Nohynkova 1978. Flagellates of the human intestine and of intestines of other species. In *Parasitic protozoa*, Vol. 2, J. P. Kreier (ed.), 2–138. New York: Academic Press.

Levine, N. D. 1985. *Veterinary protozoology*. Ames, Iowa: Iowa State University Press.

McDougold, L. R. & W. M. Reid 1978. *Histomonas meleagridis* and relatives. In *Parasitic protozoa*, Vol. 2, J. P. Kreier (ed.), 140–61. New York: Academic Press.

Meyer, E. A. & S. Radulescu 1979. *Giardia* and giardiasis. In *Advances in Parasitology* **17**, 1–47.

CHAPTER FIVE

Symbiotic amebae

The amebae are members of the subphylum Sarcodina of the phylum Sarcomastigophora. All the symbiotic forms, together with many facultatively symbiotic and free-living ones (including the well-known *Amoeba proteus*), are grouped in the superclass Rhizopoda, class Lobosea, order Amoebida, which consists of naked amebae (i.e. with a shell or "test") moving by means of lobose pseudopodia (Fig. 5.1) and never having a flagellate stage, or in the order Schizopyrenida, which may be flagellate at some stage. Only a few genera are obligate symbionts, almost always of their host's intestine, these include: *Entamoeba*, *Endolimax*, and *Iodamoeba*, all with one or more species inhabiting the gut of man; *Endamoeba* of insects, which is a name often confused with *Entamoeba*; and a few others. Only two species (both of *Entamoeba*) are known to be harmful – *E. histolytica* of man and other mammals, and *E. invadens* of reptiles (Figs 5.2 & 5.8). These sometimes invade the tissues of their hosts. Some authorities include all the obligately symbiotic amebae in a single family, the Endamoebidae, and this, while simplifying taxonomy from the parasitologist's viewpoint, suggests a closer relationship than is probably justified. It seems more likely that obligate symbiosis has been adopted independently by amebae of at least two different groups. In addition to these, other normally free-living species of amebae belonging to the families Acanthamoebidae and Vahlkampfiidae are facultative symbionts and can invade the tissues of mammals, including man, occasionally producing severe disease.

Members of the class Lobosea lack any form of sexual reproduction, reproducing only by binary (occasionally multiple) fission. Many produce resistant cysts at certain stages of their life cycle, and all are phagotrophic. The obligately symbiotic forms are transmitted directly, as far as is known, usually by ingestion of fecal material containing cysts. Almost all parasitic amebae produce cysts; an exception is *Entamoeba gingivalis*, which inhabits the mouth and upper pharynx of man. This organism is presumably transmitted directly, the trophozoite

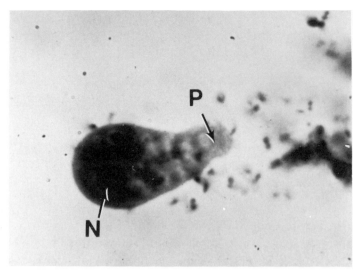

Figure 5.1 Photomicrograph of an *E. histolytica* trophozoite showing locomotion by a broad lobose pseudopodium (on the left of the organism), which is characteristic of amebae of the class Lobosea. The nucleus (N), which in this specimen is intensely stained, is in the anterior part of the organism; the uroid can be seen at the posterior end (P).

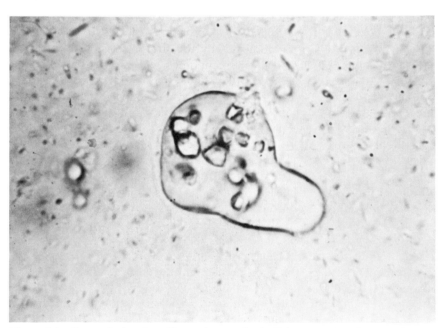

Figure 5.2 *Entamoeba invadens*. Photomicrograph of a living organism from a culture. In form and mode of locomotion it closely resembles the *E. histolytica* shown in Figure 5.1. A variety of granular inclusions is present in the ameba's cytoplasm. The organism may be pathogenic to reptiles.

(i.e. nonencysted stage) surviving since it does not have to pass through the stomach to reach its habitat.

GENUS *ENTAMOEBA* CASAGRANDI AND BARBAGALLO, 1895

This genus is distinguished by its nuclear structure (Fig. 5.3). The nucleoprotein is arranged in a peripheral ring (which appears granular in fixed specimens) lining the nuclear membrane, and in a small, more or less central, karyosome; the rest of the space within the nuclear membrane appears empty, and presumably contains fluid. All species except one are symbiotic.

The species of this genus which infect man and domestic animals can be divided into four groups, based mainly on the number of nuclei present in the mature cyst (or, in *E. gingivalis* group, the absence of a cyst). The trophozoite of all species is uninucleate when it encysts, but in many species two or three nuclear divisions soon occur so that the mature cyst contains four (Fig. 5.4) or eight (Fig. 5.5) nuclei. In any sample a small number of immature cysts with less than the normal number of nuclei may be seen, together with occasional freaks having more. The cysts often possess, when young, a vacuole containing glycogen (Fig. 5.6) (presumably a food reserve) but this has usually disappeared (i.e. been metabolized) by the time the nuclear divisions (if any) are complete. Also, cysts sometimes contain structures called chromatoid bodies (Fig. 5.7) which in fresh preparations appear glass-like but stain intensely with various hematoxylins (hence their name). The chromatoid bodies, too, disappear as the cyst ages (though less rapidly than the glycogen). They are composed of ribonucleoprotein particles arranged in a regular, almost crystalline array. As the cyst

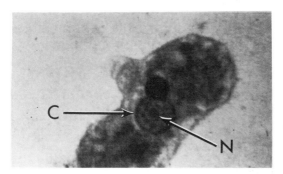

Figure 5.3 *Entamoeba histolytica*. This photomicrograph shows the appearance of the nucleus in a fixed and stained preparation. The nucleolus (N) is centrally located and much of the chromatin (C) is arranged around the nuclear periphery.

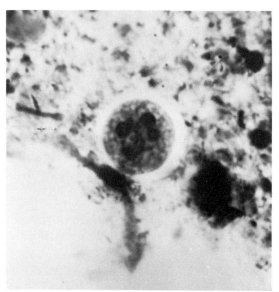

Figure 5.4 Photomicrograph showing a mature *Entamoeba histolytica* cyst which contains four deeply stained nuclei.

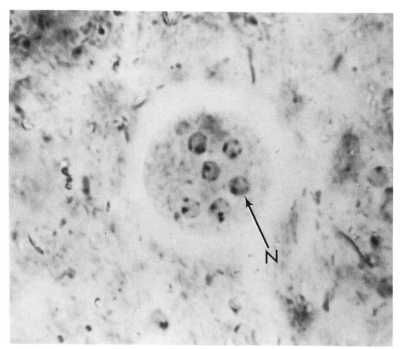

Figure 5.5 Photomicrograph of a mature *Entamoeba coli* cyst with eight nuclei (N). Two of the nuclei are barely visible as they are out of the plane of focus.

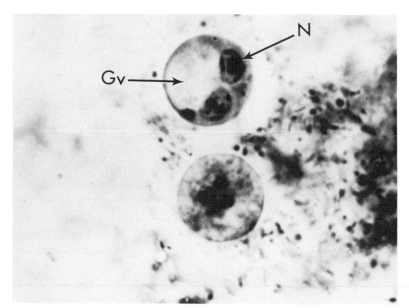

Figure 5.6 Photomicrograph of an immature, binucleate cyst of *Entamoeba coli*, containing a large glycogen vacuole (Gv). As the cyst matures the glycogen vacuole is absorbed and the nuclei (N) divide to yield a maximum of eight.

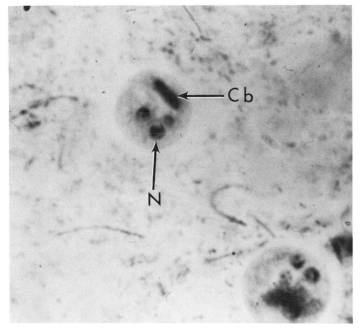

Figure 5.7 Photomicrograph showing an *Entamoeba histolytica* cyst which contains a chromatoid body (Cb) with blunt, rounded ends, and two nuclei (N).

ages the particles become dispersed throughout the cytoplasm. The shape of the chromatoid bodies is of some help in separating the species into the groups mentioned above. The arrangement is as follows.

(a) Cysts with four nuclei and chromatoid bodies which are relatively broad rods with blunt, rounded ends (Fig. 5.7). Trophozoites and cysts of this group have nuclei of rather delicate structure. Species include *E. histolytica* (man, other primates, dogs, cats, pigs, rodents), *E. hartmanni* (man and possibly the other hosts of the previous species), and *E. moshkovskii* (the only known nonsymbiotic species of this genus), which is found in sewage and was aptly described by Levine as "a parasite . . . of the municipal digestive tract." The first two species will be discussed in more detail below. *Entamoeba invadens* (reptiles), apart from *E. histolytica* the only known pathogenic species (Fig. 5.8), belongs morphologically in this group also.

(b) Cysts with eight nuclei and thin, splinter-like chromatoid bodies with pointed ends. The nuclear structure is rather coarse. Species include *E. coli* (man, other primates, dogs, possibly pigs) (Figs 5.5,

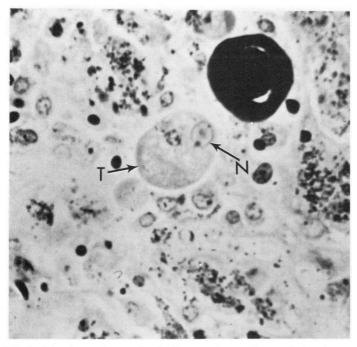

Figure 5.8 In this photomicrograph of a liver section an *Entamoeba invadens* trophozoite (T) can be seen in the tissue. The *entamoeba*-type nucleus (N) is within the plane of section.

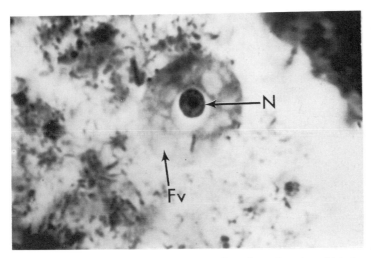

Figure 5.9 Photomicrograph showing an *Entamoeba coli* trophozoite which has coarse chromatin in an *Entamoeba*-type nucleus (N) and many large, typical food vacuoles (Fv).

5.6 & 5.9), *E. muris* (rats, mice), and *E. gallinarum* (chickens, etc.).

(c) Cysts with one nucleus: nuclear structure and chromatoid structure are variable, but the chromatoids are usually of the type seen in group (a). Species include *E. bovis* (cattle), *E. ovis* (sheep), *E. suis* (pigs), and *E. chattoni* (monkeys). No species of this group normally infects man; occasional reports of human infection with a species of *Entamoeba* producing uninucleate cysts (sometimes called "*E. polecki*") probably represent sporadic or spurious infections of man with *E. suis*. Spurious infections are those in which a resistant stage of a protozoan passes unchanged through the alimentary canal of another animal without undergoing development. Definite instances of this occurring in man are known among the coccidia.

(d) Species not producing cysts. These organisms all live in the buccal cavity of their hosts. They include *E. gingivalis* of man, other primates, dogs and cats, and similar species of horses and pigs. Many other species infecting vertebrates and invertebrates are known.

Entamoeba histolytica

The important species in the genus *Entamoeba* is *E. histolytica*. The major host of this species is man. It has been reported throughout the world, though many records probably refer to *E. hartmanni*. *Entamoeba histolytica* is sometimes pathogenic to man and its other hosts, causing amebic dysentery or amebiasis.

Morphology and life cycle   When alive and healthy, trophozoites of *Entamoeba histolytica* (Fig. 5.1) move with a forward flowing movement, rather reminiscent of a garden slug (*Limax* sp.) and hence sometimes called "limax-type movement." The entire body, elongated and of fairly regular outline, flows smoothly forward, all of its anterior end functioning as a single, broad pseudopodium. It is only at this anterior pole that the outer clear ectoplasm is sharply differentiated from the inner granular endoplasm. When kept on a microscope slide at a temperature below 37 °C, the ameba becomes unhappy and moves in a far less single-minded fashion, putting out tentative pseudopodia in various directions and withdrawing them again (Fig. 5.10). It does not survive long under such conditions, but soon rounds up and dies. In this state, which is how it is usually seen in stained preparations on microscope slides, the distinction between ectoplasm and endoplasm is more clearly visible. The trophozoite contains a single nucleus of the cartwheel type with, in stained preparations, a rather fine central karyosome and peripheral granules. Food vacuoles containing various objects such as bacteria, host-cell nuclei and, if the infection is pathogenic, host erythrocytes may be seen in the endoplasm. There is no contractile vacuole (Fig. 5.11). Trophozoites are found in the host's large intestine, usually in the lumen where they are 10–20 μm in diameter. Sometimes, for unknown reasons, trophozoites invade the mucosa and submucosa, and may spread to other tissues, chiefly the liver. In these richer environments, they become larger (20–50 μm in diameter).

The trophozoites encyst only in the gut lumen, not in the tissues. They round up and become quiescent, and secrete a thin cyst wall. A

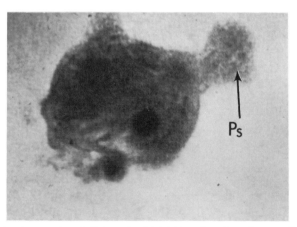

Figure 5.10 Photomicrograph of *Entamoeba histolytica* showing the small irregular pseudopodia (Ps) formed by the ameba under the poor conditions which often occur on a microscope slide.

114

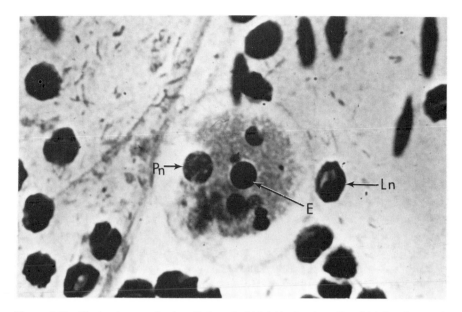

Figure 5.11 Photomicrograph of an *Entamoeba histolytica* trophozoite which has ingested erythrocytes (E). The parasite nucleus (Pn) and nuclei of leukocyte (Ln) in the tissue surrounding the amebae are visible.

diffuse mass of glycogen may be seen in young cysts, but this soon disappears. Chromatoid bodies are usually seen in young cysts, but they disperse as the cyst ages. The single nucleus of the trophozoite soon undergoes two divisions to produce the characteristic four nuclei of the mature cyst, which on the average is about 12 μm in diameter (range 9.5–15.5 μm). The cyst is the transmissive stage; it does not develop further in the original host, but is passed out in the feces. It can survive outside the host for some time, and when swallowed by another susceptible mammal passes unharmed through the stomach. The parasite probably emerges from the cyst (excysts) in the small intestine as a four-nucleate ameba; its nuclei divide once more, and then cytoplasmic division occurs, producing eight small uninucleate amebae which pass to the large intestine and grow to full size. Here (and in the tissues, if invasion occurs) the amebae multiply by binary fission.

Epidemiology   Infection with *E. histolytica* depends on ingestion of fecal material containing cysts. Since few people are coprophagous, ingestion is usually with contaminated food or water. The source of infection is usually man, for though infected dogs or rats may occasionally pass the parasite to man it is probably more usual for them to become infected from us. Thus, the infection is commoner in areas where food hygiene is poor. Possible sources of infection are food-handlers who are themselves chronically infected without being ill ("cyst-passers") and whose standards of cleanliness are not ideal. The use of untreated human feces as a fertilizer is a possible source of infection, as is contamination of water supplies and their inadequate purification. Coprophagous insects such as flies and cockroaches may carry viable cysts to foodstuffs, on their legs, in their probosces, or in their intestines. The cysts will survive for several weeks outside the body if not desiccated. They are not killed by aqueous potassium permanganate, a "disinfectant" often used for washing salads, etc. in tropical countries but are readily killed by heat (50 °C or above).

Pathogenesis *Entamoeba histolytica* usually lives in the gut lumen as a harmless commensal. In some infections, however, the trophozoites penetrate the mucosa and the muscularis mucosae and invade the submucosa (Fig. 5.12). The reasons why some infections are pathogenic and others are not are still unclear. It has been suggested that pathogenicity may be associated with a particular bacterial flora in the individual's intestine, but more recent work has indicated that there

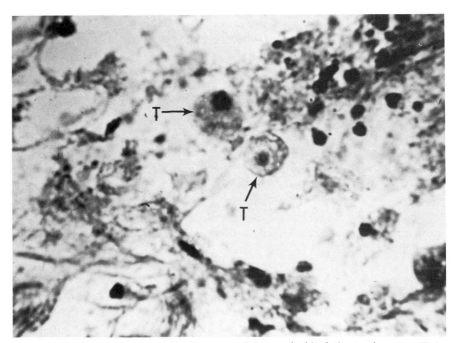

Figure 5.12 Photomicrograph showing several *Entamoeba histolytica* trophozoites (T) in the submucosa. The tissue is edematous and badly damaged.

may be a particular zymodeme (i.e. a population defined by the characteristics of certain enzymes) of *E. histolytica* which is, or tends to be, pathogenic. Whatever the causation may be, once the amebae have reached the submucosa they multiply and spread radially outwards below the mucosa to form a characteristically flask-shaped lesion or ulcer, the centre of which is filled with cellular debris, lymphocytes, plasma cells, and macrophages. Secondary bacterial infection may occur and then polymorphonuclear leukocytes will also be found. As the submucosa is eroded by the amebae (which are found chiefly at the advancing edges of the ulcer), many blood vessels are broken and the typical bloody dysentery results. Rarely, the ulcers perforate the gut wall entirely and cause peritonitis. A more common complication is the spread of amebae via the blood vessels to other organs, where they also invade and destroy tissue causing amebic abscesses (in which bacteria may not be present). The abscesses may become very large – several centimeters in diameter. The commonest site for their development is the liver, because most of the blood from the gut is carried there by the hepatic portal system; but they may be found in almost any other organ, particularly the lungs and brain.

Untreated amebic dysentery may result in death from loss of fluid and blood. If abscesses develop in the liver or elsewhere, apart from the damage done to the organ concerned, the abscess may finally rupture either through the body wall or into the peritoneal cavity with serious, sometimes fatal, consequences.

Diagnosis Amebiasis is usually diagnosed by recovery of the protozoa from the feces. If dysentery is present, active trophozoites may be seen in fresh feces mixed with saline and examined microscopically at 37 °C. (It is, however, possible for trophozoites of other nonpathogenic ameba to be passed out if diarrhea is present due to some other cause, so any trophozoites seen must be identified morphologically.) In diarrheic and normal feces, cysts may be found either by direct microscopical examination or after their concentration from the sample. Finally, in cases of doubt, attempts may be made to cultivate the amebae in suitable artificial media. Serological methods of diagnosis have been developed, but generally detect only cases when tissue invasion has occurred.

Treatment and prevention Various drugs are available to control infection with *E. histolytica* although complete cure is not always easy. The alkaloid emetine has been used for many years (in various compounds) and is effective though rather toxic. A synthetic derivative, dehydroemetine, is equally effective and less toxic; both are given intramuscularly. Metronidazole or tinidazole are now commonly used,

117

often with tetracycline. The antimalarial drug chloroquine is effective against amebic abscesses in the liver but not elsewhere. Diloxanide, which destroys amebae only in the gut lumen, may be used to treat symptomless "carriers" or as a prophylactic. Large abscesses may have to be drained surgically.

Prevention of infection is entirely a matter of hygiene, both personal (washing of hands, avoiding the eating of raw vegetables and salads in dangerous areas, protection of food from coprophilic insects, etc.) and municipal (sewage disposal and water purification). If possible, food handlers in endemic areas should be examined for infection and treated if necessary (a counsel of perfection which is seldom practicable).

OTHER INTESTINAL AMEBAE OF MAN

In addition to *E. histolytica* four other species of amebae inhabit the human intestine (Table 5.1). All are nonpathogenic, but when attempting to diagnose amebiasis it is important to be aware of their presence to avoid possible confusion. *Entamoeba hartmanni* closely resembles *E. histolytica*. The trophozoites are about the same size as the smaller individuals of *E. histolytica* (9–14 μm) while the cysts, which have four nuclei and chromatoid bodies exactly like those of *E. histolytica*, are definitely smaller (4.0–10.5 μm in diameter, average 7.4 μm). Thus, *Entamoeba*-type cysts in human feces which are less than 10 μm in diameter are almost certainly not *E. histolytica*.

Much confusion surrounds the taxonomy of *E. histolytica* and *E. hartmanni*, the latter being often referred to as the "small race" of the former, but the evidence that they are separate species seems to be good. It is probable that many records of *E. histolytica* indigenous to the temperate parts of the world refer to *E. hartmanni*.

The main distinguishing features of the five species of human intestinal amebae are listed in Table 5.1. *Endolimax nana* (Fig. 5.13) does not have a cartwheel-type nucleus and *Iodamoeba buetschlii* (Fig. 5.14) has a distinct glycogen vacuole in its cyst.

GENERA *ACANTHAMOEBA* VOLKONSKY, 1931 AND *NAEGLERIA* ALEXEIEFF, 1912

In recent years there has been an increasing number of reports of human infections by amebae which are normally free-living inhabitants of soil or water. The most serious are those due to *Naegleria fowleri* (subclass Gymnamoebia, order Schizopyrenida); another species, *N.*

Table 5.1 Intestinal amebae of man.

	Trophozoite			Cyst			
	size (μm)*	nuclear structure	size (μm)*	no. of nuclei (mature)	chromatoid bodies	special features
Entamoeba histolytica	10–40	Entamoeba-type, delicate	9.5–15.5 (~ 12)	4	Broad, blunt ends	—
E. hartmanni	9–14	Entamoeba-type, coarser	4–10.5 (~ 7.4)	4		—
E. coli	15–30	Entamoeba-type, coarser	10–30 (~ 17)	8	Thin, sharp ends	—
Endolimax nana	6–12 (~ 3)	Not Entamoeba-type	6–9 × 5–7	4	None	—
Iodamoeba buetschlii	5–20 (~ 11)	Not Entamoeba-type	9–15	1	None	Persistent glycogen vacuole†

* Means given in parenthesis.
† Stains golden-brown color with iodine solution.

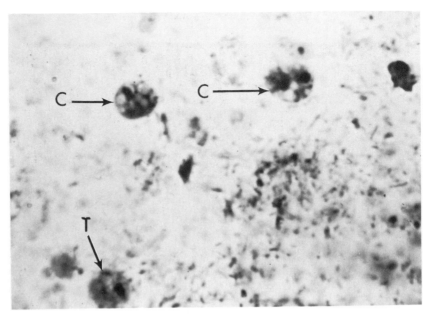

Figure 5.13 Photomicrograph of *Endolimax nana* trophozoite (T) and cysts (C) in a fecal preparation.

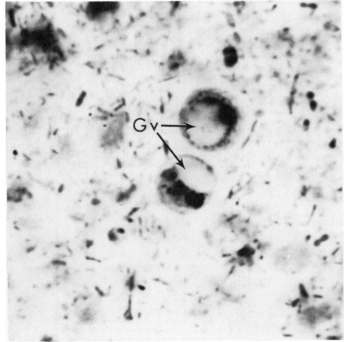

Figure 5.14 Photomicrograph of *Iodamoeba buetschlii* cysts showing large glycogen vacuoles (Gv) which are typical of the cysts of this organism.

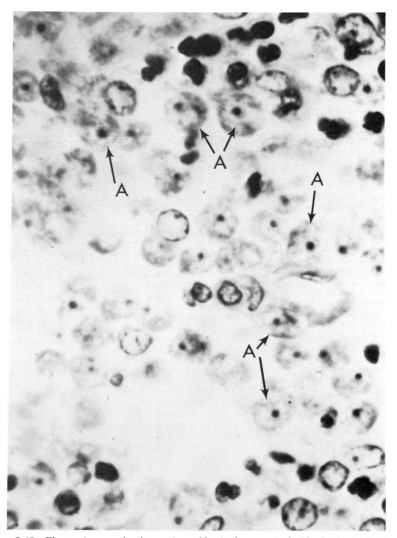

Figure 5.15 Photomicrograph of a section of brain from an individual who had primary amebic meningoencephalitis caused by *Naegleria*. Amebae (A) are present throughout the tissue.

australiensis, can be pathogenic, at least in experimentally infected mice. Amebae of the genus *Naegleria* have flagellate and cyst stages in their life cycles. The flagellate stages occur commonly in warm water. Some of the infected persons were known to have been swimming in muddy water and amebae probably entered their mouths or noses. The amebae migrate up the olfactory tract to the brain and cause a rapidly fatal disease called primary amebic meningoencephalitis, or PAM

(Fig. 5.15). The rapidity of development of this disease makes treatment difficult if not impossible, but, if started early enough, amphotericin B given together with sulphonamides or tetracyclines, or with rifampicin plus miconazole, has had some successes.

A more chronic, granulomatous encephalitis has been, rarely, found to be caused by *Acanthamoeba culbertsoni* (order Amoebida), especially in immunodeficient individuals. No successful treatment is known. Other species of *Acanthamoeba* have been known to cause corneal ulcers, which may respond to treatment with diamidines and neomycin, or to natamycin (an antifungal antibiotic).

Symptomless infections of the throat of man with these amebae have also been found in the USA. As most of the infections were in infants at the crawling stage, the likeliest means of infection was the ingestion of soil or dust containing cysts or trophozoites of the amebae. It is possible that, in hosts whose immune mechanisms are impaired, these normally commensal amebae may become invasive and lead to meningitis.

FURTHER READING

Albach, R. A. & T. Booden 1978. Amoebae. In *Parasitic protozoa*, Vol. 2, J. P. Kreier (ed.), 455–507. New York: Academic Press.

Ash, L. R. & T. C. Orihel 1980. *Atlas of human parasitology*. Chicago: American Society of Clinical Pathologists.

Culbertson, C. G. 1971. The pathogenicity of soil amebas. *Annual Review of microbiology* **25**, 231–54.

Elsdon-Dew, R. 1968. The epidemiology of amoebiasis. *Advances in Parasitology* **6**, 1–62.

John, D. T. 1982. Primary amebic meningoencephalitis and the biology of *Naegleria fowleri*. *Annual Review of Microbiology*, **36**, 101–23.

Martinez, A. J. 1985. Free-living amebas: natural history, prevention, diagnosis pathology, and treatment of diseases. Boca Raton, Florida: CRC Press.

Martinez-Palomo, A. 1982. *The biology of* Entamoeba histolytica. New York: Research Studies Press & Wiley.

CHAPTER SIX

Gregarines, hemogregarines, and intestinal coccidia

This is the first of three chapters dealing with the phylum Apicomplexa (previously named Sporozoa), all members of which are parasitic.

The Apicomplexa probably all evolved from a single ancestor. The life cycles have obvious similarities and the motile stages all possess the characteristic apical complex visible with the aid of electron microscopy. Only one (Sporozoea) of the two classes (Perkinsea and Sporozoea) contains organisms of medical or veterinary importance. Some of these, the malaria parasites and piroplasms, will be considered in the next two chapters.

Many Apicomplexa possess spores, originally a resistant stage involved in dispersion and transmission. In more sophisticated members of the subphylum (e.g. the Haemosporina), the spore as a resting stage with a resistant wall may be lacking since, with the adoption of an insect vector, it is no longer necessary to protect the organisms while outside a host. In the Haemosporina, the infective stage (sporozoite) which develops within the spore of the more primitive members can be recognized. However, in the piroplasms not even vestiges of the spore have been identified, and many unrelated organisms also produce spores; this is why the phylum was redefined with possession of the apical complex as the major diagnostic feature.

CLASS SPOROZOEA

Members of this class reproduce both sexually and asexually. They constitute the majority of species of Apicomplexa and include the causative organisms of malaria.

Probably all Apicomplexa are haploid for almost all of their life cycles, meiosis occurring in the first zygotic nuclear division (i.e. the first division following fertilization). Thus only the zygote is diploid.

123

There are two subclasses in the Sporozoea, the Gregarinia and the Coccidia.

SUBCLASS GREGARINIA

The gregarines are all parasites of the gut and the hemocoel of invertebrates, mainly insects. For most of their life cycles they are extracellular. They have no vectors, transmission being by ingestion of the spores liberated from an infected host. The sporozoites released from the swallowed spores enter cells of the host (often those of the gut wall) where they increase in size. In most gregarines (the order Eugregarinida) the trophozoites ("gregarins") emerge from the host cell, but remain at first attached to it by the anterior end (Fig. 6.1). In many eugregarines (the cephaline eugregarines) the body is divided

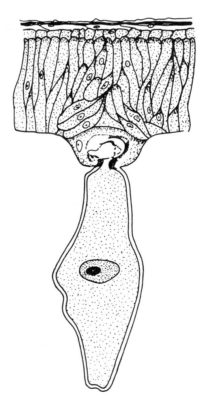

Figure 6.1 A gregarine (*Lecudina fluctus*) from the oligochaete worm *Urechus unicinctus*, shown attached to the lining of the mid-intestine. Note the epimerite imbedded in an epithelial cell, apparently causing some hypertrophy (after Iitsuka, from R. D. Manwell 1961).

into two compartments, an anterior protomerite and a posterior deutomerite (Fig. 6.2); on the front of the protomerite is a protrusion called the epimerite, by means of which the trophozoite is attached to the host cell. These two compartments should not be regarded as separate cells since each trophozoite has only a single nucleus, usually in the deutomerite. In other (acephaline) eugregarines, this division of the body is not seen. The parasite *Monocystis* is an acephaline eugregarine.

The trophozoites of most gregarines eventually break away from the host cell and move in the hemocoel, gut lumen or other body cavity of

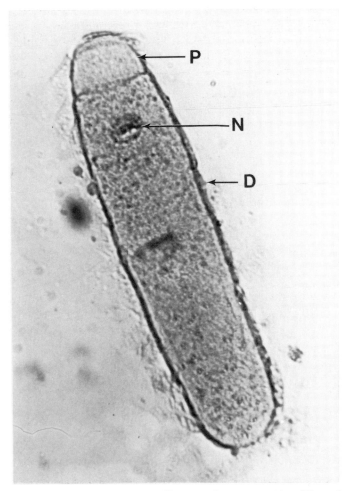

Figure 6.2 A cephaline eugregarine. The anterior protomerite (P) and posterior deutomerite (D) compartments of the cell are clearly visible. The nucleus (N) is in the deutomerite.

the host. In a minority of gregarines, the trophozoites remain intracellular and reproduce by merogony. Many such cycles of merogony may occur and each time a meront matures one host cell is destroyed and many more are invaded by the released parasites (merozoites). Eventually, meronts (often of a different type, with larger nuclei) are formed; the merozoites resulting from these meronts remain extracellular and grow into sexual individuals or gametocytes ("sporadins"). In the gregarines without merogony all the trophozoites become gametocytes. Two gametocytes then become attached to one another and after a longer or shorter interval they encyst. This association of gametocytes (presumably of opposite sexes) is called syzygy (Fig. 6.3). The associated pair within the cyst (gametocyst) undergoes nuclear division and buds off small (usually identical) individuals (gametes) from the surface. The gametes (which do not possess flagella) fuse in pairs to form zygotes, each of which then encysts within an oocyst (or "spore"). Many zygotes are formed by the Eugregarinida, but fewer (often only two) by the gregarines which undergo merogony. The oocysts are at first within the gametocyst but later, either before or after leaving the host, they escape from it. The contents of each sporocyst divide to form several (usually eight) sporozoites. The sporocysts are liberated in some way from the body of the host and, by virtue of their cyst wall, are resistant to desiccation and the other hazards of life outside a host. The sporocyst is the transmissive stage and after it has been swallowed by a new host, the sporozoites emerge from it and the life cycle begins again (Fig. 6.4).

Generally speaking, the gregarines which do not undergo merogony are harmless to their hosts, while the others, since they increase greatly in number by repeated merogonies within the host, are usually pathogenic. Classification of the gregarines is based mainly on the presence or absence of merogony. Following Grassé, the Society of Protozoologists considered that the forms which undergo merogony are not all closely related, and so grouped them in two separate orders:

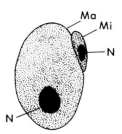

Figure 6.3 *Adelea* sp.; macro- and microgametocytes (Ma, Mi) in syzygy (host unknown); N = nucleus.

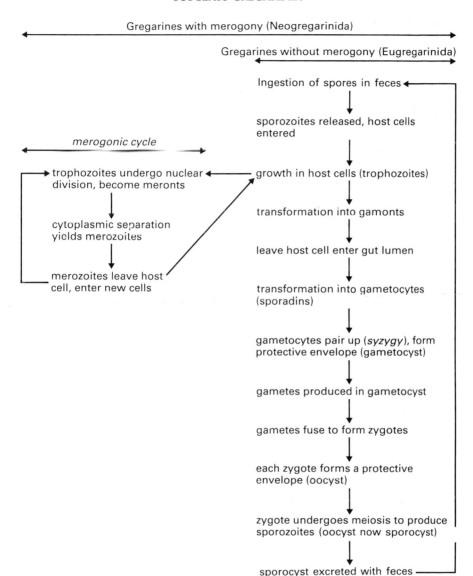

Figure 6.4 Generalized life cycle of gregarines. In those with merogony the numbers of organisms are increased by cycles of asexual reproduction.

Archigregarinida (found mainly in marine annelids) and Neogregarinida (in insects). The forms which lack merogony (the majority) are placed in the order Eugregarinida (found in annelids and arthropods). It is thought that the Archigregarinida are a primitive group, the Eugregarinida have evolved later and have lost merogony, and the

127

Neogregarinida are the most recently developed group, which have re-acquired merogony. This idea seems inherently improbable. Kudo groups all the gregarines which lack merogony in the order Eugregarinida and those which possess it, in the order Neogregarinida.

SUBCLASS COCCIDIA

Of the three orders within this class, the first two, Agamococcidiida and Protococcidiida, are represented by only a few species, mainly parasitic in marine annelids. The third order, Eucoccidiida, contains three suborders: Adeleina, Eimeriina and Haemosporina (see Ch. 7).

ORDER EUCOCCIDIIDA, SUBORDER ADELEINA

The Adeleina are distinguished by the fact that males and females develop in association with each other (syzygy, as in gregarines). Many members of this suborder which live in red or white blood cells, and sometimes other cells, of vertebrates (of all classes) are called "hemogregarines," an imprecise term which also includes some Eimeriina. These are transmitted from one vertebrate to another by invertebrate vectors (leeches, insects, ticks, and mites), in which they undergo sexual reproduction; asexual multiplication (merogony) occurs in the vertebrate host. The adeleine hemogregarines can be distinguished from malaria parasites (which also inhabit erythrocytes) by the fact that they contain no pigment. They are grouped in three families: Haemogregarinidae, Hepatozoidae and Karyolysidae (containing the genera *Haemogregarina* (Figs 6.5, 6.6), *Hepatozoon* (Fig. 6.7), and *Karyolysus* respectively). After merogony in the vertebrate, the sexual precursors (gametocytes) are produced. These always inhabit blood cells, in which they are sucked up by the next blood-sucking invertebrate to feed on their host. If the invertebrate is of the correct species, the gametocytes associate in syzygy and produce sexual individuals (gametes). Fertilization occurs and the zygote, which in the Hepatozoidae is motile and is called an ookinete, usually encysts to form an oocyst in the lumen or wall of the vector's gut or in its hemocoel. In some Karyolysidae, the nonmotile zygote divides into several motile sporokinetes which invade the vector's ovaries and encyst in the eggs to form sporocysts. The oocyst (or sporocyst) grows and its contents undergo repeated nuclear division and eventually differentiate into several (or many) small, elongate, uninucleate sporozoites (this process being called sporogony). The sporozoites, by

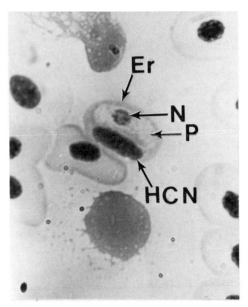

Figure 6.5 In this photomicrograph a hemogregarine can be seen lying within an erythrocyte of a frog. The host cell nucleus (HCN) is pushed to one side of the erythrocyte (Er) by the development of the protozoan (P). The protozoan nucleus (N) is centrally located.

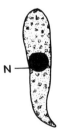

Figure 6.6 *Haemogregarina* sp. free in the blood of a frog; N = nucleus.

Figure 6.7 *Hepatozoon balfouri* (P) in erythrocyte (Er) of jerboa. The nucleus (N) of this organism is clearly visible.

various routes, enter the next vertebrate on which the vector feeds; certainly in some species, and perhaps in all, they enter through the mucous membranes if the vertebrate eats the vector or, in some instances, its feces, which contain sporocysts. The route of entry of most species is unknown. If the vertebrate is susceptible to the parasite, the sporozoites enter the appropriate cells and commence merogony. Eventually, gametocytes are produced and the cycle is complete (Fig. 6.8).

This life cycle has many similarities to that of the coccidia and the malaria parasites, except that syzygy does not occur in the latter and the former have only one host. Its complexity may appear rather alarming at first sight, but basically it may be reduced to the following components: merogony (asexual multiplication) in the vertebrate host, gametogony (development of the sexual gametes) begun in the vertebrate and completed in the invertebrate; followed by fertilization and sporogony (multiplication immediately following fertilization), both in the invertebrate host.

As far as is known, most adeleine hemogregarines are normally nonpathogenic; some species of the genus *Hepatozoon* may, however, cause disease in their hosts (e.g. *H. muris*, in mice). The remaining Adeleina, which do not inhabit blood cells and complete their life cycle in a single host (vertebrate or invertebrate), are contained in the two families Adeleidae and Klossiellidae; the oocyst is passed out in the host's feces or urine in a manner similar to that of the Eimeriidae.

SUBORDER EIMERIINA

This group includes the organisms often colloquially referred to as "coccidia". A few of the Eimeriina, in which certain stages inhabit the blood cells of vertebrates, are included in the loose term "hemogregarines". The eimeriine hemogregarines are distinguished from the adeleine hemogregarines by the absence of syzygy.

Two important genera in the Eimeriina are *Eimeria* and *Isospora*. Species of the genus *Eimeria* have only a single host in which sexual and asexual reproduction occur. The oocyst resulting from the sexual cycle is a thick-walled resistant stage which can survive for long periods outside the host, and is the transmissive stage. Thus, *Eimeria* can be regarded as monoxenous, i.e. having only one host in its life cycle. Eimeriina of the genera *Toxoplasma*, *Sarcocystis*, *Frenkelia* and *Besnoitia*, in contrast, have one final or definitive host, in which sexual and asexual reproduction occurs, as with *Eimeria*, but they are also able to infect a variety of other, intermediate, hosts in which only asexual reproduction occurs. Transmission may result from ingestion of either

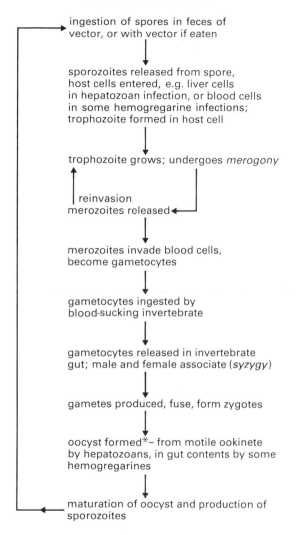

ingestion of spores in feces of
vector, or with vector if eaten

sporozoites released from spore,
host cells entered, e.g. liver cells
in hepatozoan infection, or blood cells
in some hemogregarine infections;
trophozoite formed in host cell

trophozoite grows; undergoes *merogony*

reinvasion
merozoites released

merozoites invade blood cells,
become gametocytes

gametocytes ingested by
blood-sucking invertebrate

gametocytes released in invertebrate
gut; male and female associate (*syzygy*)

gametes produced, fuse, form zygotes

oocyst formed*– from motile ookinete
by hepatozoans, in gut contents by some
hemogregarines

maturation of oocyst and production of
sporozoites

* In some hemogregarines (e.g. leech-transmitted forms) no oocyst
is formed, sporozoites are formed from the zygote and make their way
to the salivary glands, from which the next host is infected when the
vector feeds.

Figure 6.8   Generalized life cycle of adeleine hemogregarines.

infected carcasses of intermediate hosts or oocysts excreted by final
hosts. In some of the isosporan Eimeriina alternation of these two
types of host in the life cycle is obligatory; in others (chiefly the genus
Toxoplasma) it is not, as transmission can occur directly from one
intermediate host to another. Thus species of these genera, unlike
Eimeria, are heteroxenous, some obligatorily and some facultatively.

Isospora, as classically defined, is also a monoxenous genera but, as pointed out below, it now seems probable that many if not all species of *Isospora* may have two hosts, like *Toxoplasma*.

The sexual development of Eimeriina usually occurs in cells of the intestinal epithelium, though in some species it occurs in epithelium lining ducts in the liver or kidneys. These locations permit the oocyst, the product of sexual development, to be passed with the urine or feces. Infection is usually by ingestion of contaminated food or water. Vertebrates and invertebrates, including molluscs and arthropods, are hosts for coccidia.

Figure 6.9 *Lankesterella garnhami* in a monocyte in the spleen of a sparrow. The protozoan (P) lies just below the host cell nucleus (HCN).

The eimeriine hemogregarines all belong to the genera *Lankesterella* or *Schellackia* (Fig. 6.9). They are common in birds, reptiles, and amphibia.

The classification of the Eimeriina is in a state of flux. The classic version is based almost solely on the number of sporocysts and sporozoites within the oocyst. By this system the Eimeriina were grouped into 25 (now 28) genera in eight subfamilies (Table 6.1). While this system is probably too artificial, being based on only one criterion, there is at present no consensus on alternative taxonomies. The relatively recent discovery that at least some species of *Toxoplasma*, *Sarcocystis* and *Besnoitia* have an "isosporan" phase in their life cycles has led some workers to regard all these names as synonyms of *Isospora*; others have resisted this view. For the sake of simplicity, and to avoid entry into a rapidly developing and highly controversial area, we have retained the "old" generic names. Species of medical or veterinary importance assigned to the genera *Eimeria*, *Isospora*, *Toxoplasma*, *Sarcocystis*, *Besnoitia*, and *Cryptosporidium* will be considered in more detail in the remainder of this chapter. Another genus, *Hammondia*, which was erected within this suborder, is almost certainly a synonym of *Toxoplasma*.

GENUS *EIMERIA* SCHNEIDER, 1895

This genus contains a large number of species infecting vertebrate animals of all classes. So far, only a single host species is known for

Table 6.1 Subfamilies and genera of the Eimeriidae (after Pellérdy 1965).

Subfamily	Genus	Number of	
		Sporocysts	Sporozoites*
Cryptosporidunae	Cryptosporidium	0	4
	Pfeifferinella	0	8
	Schellackia	0	8
	Tyzzeria	0	8
	Lankesterella	0	∞
Caryosporinae	Mantonella	1	4
	Caryospora	1	8
Cyclosporinae	Cyclospora	2	2
	Isospora	2	4
	Toxoplasma†	2	4
	Sarcocystis†	2	4
	Besnoitia†	2	4
	Dorisiella	2	8
Eimeriinae	Eimeria	4	2
	Wenyonella	4	4
	Angeiocystis	4	8
Yakimovellinae	Octosporella	8	2
	Yakimovella	8	∞
Pythonellinae	Hoarella	16	2
	Pythonella	16	4
Barrouxinae	Barrouxia	∞	1
	Echinospora	∞	1
Aggregatinae	Merocystis	∞	2
	Pseudoklossia	∞	2
	Aggregata	∞	3
	Caryotropha	∞	12
	Myriospora	∞	∞
	Ovivora	∞	∞

* The number of sporozoites is that in each sporocyst, except in the Cryptosporidiinae, when the total number per oocyst is given.

† These genera have been fitted into this taxonomic scheme only since the discovery of their sexual cycle. They are treated by some authorities as synonyms of Isospora, while others regard them as replacements for Isospora (see text).

each species of *Eimeria*. A few species have been recorded, some rather doubtfully, from annelids, arthropods (mostly centipedes) and proto-chordates. None infects man, though oocysts of one or two species which infect fish have been seen in the feces of persons who have eaten infected fish; these have been erroneously described as "parasites" of man ("spurious infections"). Species of *Eimeria* are important pathogens of chickens and calves, as well as other domestic livestock (Table 6.2).

Morphology and life cycle

The life cycle of *Eimeria* is typical of monoxenous coccidia (Fig. 6.10) An oocyst (Fig. 6.11), when swallowed, hatches in the small intestine of the host (probably under the influence of mechanical pressure, pepsin and trypsin). The sporozoites, which emerge from the sporocysts and penetrate the cells of the intestinal mucosa, may round up and grow in these cells, or they may be carried elsewhere in the body by macrophages, depending on the species. The growing forms are called trophozoites (Fig. 6.12). Most of them begin nuclear division, thus becoming meronts (Fig. 6.13). Cytoplasmic division then occurs to

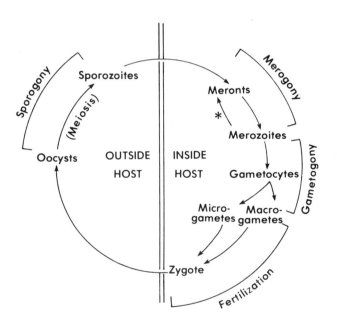

* Not in all species

Figure 6.10 Diagrammatic representation of the life cycle of a monoxenous eimeriine coccidian.

Table 6.2 Some of the more important species of *Eimeria* which infect domestic animals.

Host	Species	Habitat	Pathogenicity
cattle	E. bovis	small & large intestine	fairly high
	E. zurnii (+17 other species)	small & large intestine	fairly high
sheep and goats	E. ahsata	intestine	moderate
	E. arloingi (+8 other species)	small intestine	moderate
pigs	E. debliecki (+ 5 other species)	small & large intestine	mild–moderate
chickens	E. necatrix	small intestine & ceca	high
	E. tenella (+ 6 other species)	ceca	high
turkeys	E. adenoeides	small & large intestine	moderate
	E. meleagrimitis (+5 other species)	small intestine	moderate
ducks	E. bucephalae (+1 other species*)	small intestine	high
geese	E. truncata (+5 other species)	kidney tubules	high
rabbits	E. stiedai	bile duct of liver	high
	E. irresidua	small intestine	moderate
	E. magna (+2 or more other species)	small intestine	moderate
dogs and cats	E. canis (dogs only) (1 or 2 other species of *Eimeria* in cats only)	intestine	moderate

* A species of another genus, *Tyzzeria perniciosa*, is pathogenic to ducks.

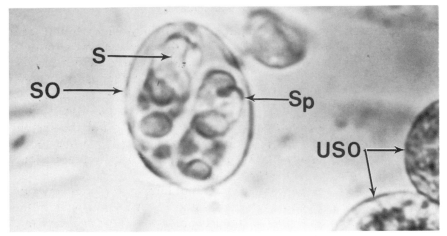

Figure 6.11 *Eimeria acervulina*, oocysts. A sporulated oocyst (SO) containing sporocysts (Sp) and sporozoites (S) is visible in the center of this photomicrograph, while portions of two unsporulated oocysts (USO) are visible in the lower right.

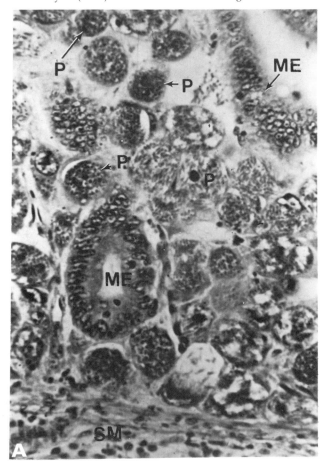

produce the small (usually 5–15 μm long by 1–2 μm wide), uninucleate organisms called merozoites. (Note that this process, merogony, is asexual.) The size of the meronts varies widely with different species, ranging up to about 10 μm in diameter. The number of merozoites produced varies from about 16 up to thousands. These quantities may also vary at different stages in the life cycle of the same species. The merozoites are liberated from the host cell when it and the meront rupture. In most species they re-enter other cells (either nearby or, in some species, in more distant tissues) to recommence merogony for a varying, but limited, number of generations.

Sooner or later, for reasons which are not known, the merozoites do not recommence merogony but instead enter host cells (usually of the intestinal mucosa) and develop by a process called gametogony into sexual individuals, the gametocytes (Fig. 6.14). These grow; the female,

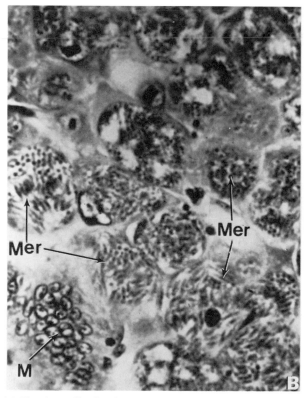

Figure 6.12 (a) *Eimeria tenella,* developing protozoa (P) are present in the cecal mucosa of a chicken. The mucosal epithelium (ME) and submucosal tissue (SM) are relatively undamaged. (b) Meronts in advanced stages of development are visible in this micrograph. The merozoites (Mer) appear either as circular objects or as long structures depending on the plane of section.

137

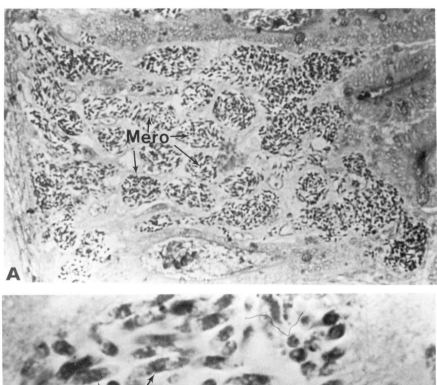

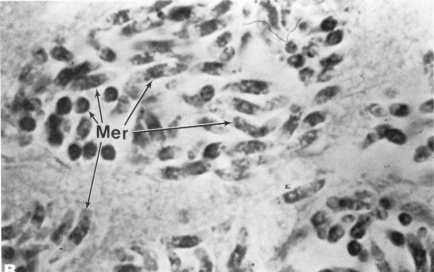

Figure 6.13 *Eimeria tenella* merozoites in mature meronts in cecal tissue; (a) low magnification showing the nests of meronts (Mero) in the tissue, and (b) high magnification showing the merozoites (Mer).

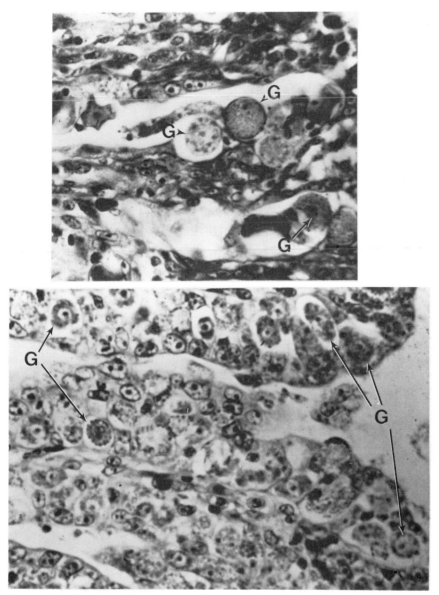

Figure 6.14 *Eimeria tenella* gametocytes in cecum of a chicken. Numerous gametocytes (G) in various stages of development are present in these sections of cecal mucosa.

or macrogametocyte, remains uninuclear, but the male, or microgametocyte, undergoes repeated nuclear division and finally produces at its surface a large number of small (2–3 μm long) curved organisms consisting mainly of nucleus, mitochondrion and two flagella. These small organisms are the motile male gametes, or microgametes. Each swims away in search of a female and, on finding one, fertilizes her. The fertilized female, now called a zygote, encysts (still within the host cell). The immature oocyst leaves the host cell and, in the forms inhabiting the intestine, is passed with the feces. If the conditions of moisture, temperature, and oxygen tension are correct sporulation will occur. Within the thick cyst wall, the zygote's nucleus (now diploid as a result of fertilization) divides meiotically; cytoplasmic division follows, producing four cells called sporoblasts. Each sporoblast then encysts (within the oocyst) to form a sporocyst, the contents (nucleus and cytoplasm) which divide again to produce the two sporozoites which are contained in the sporocyst (Fig. 6.11). Often some residual cytoplasm is left unused after either or both of these divisions.

In the outside world, the sporulated oocyst can survive for a considerable time until eaten by another susceptible host and thus be enabled to re-start the complicated life cycle (Fig. 6.10).

Since the first post-fertilization nuclear division is a meiosis, it is clear that species of *Eimeria* (and indeed, as far as is known, all Eimeriina, as well as Haemosporina, and probably Adeleina) are haploid for almost all of their life cycle. This development cycle has many similarities with that of the Haemosporina and the adeleine hemogregarines, except that in the latter two groups fertilization and sporogony occur in a second, vector, host, and at no time is the parasite exposed to the dangers of the outside world.

GENUS *ISOSPORA* SCHNEIDER, 1881

Coccidia of the genus *Isospora*, defined by the classic criterion which is morphology of the oocyst, were those coccidia whose oocysts contained two sporocysts, each containing four sporozoites (Fig. 6.15). Members of well-established genera such as *Toxoplasma* and *Sarcocystis* have been demonstrated to have hosts in which a sexual cycle occurs, producing isosporan-type oocysts. In some instances both the asexual forms in the alternate host, and the sexual forms in the definitive host, were known, but their relationship was not recognized. For example, the sexual stages of a coccidian causing diarrhea in man were named *Isospora hominis* and the asexual stages of the same parasite in the muscles of bovines were named *Sarcocystis cruzi*. In the same fashion, a coccidian of the buzzard was designated *Isospora buteonis* and its stages

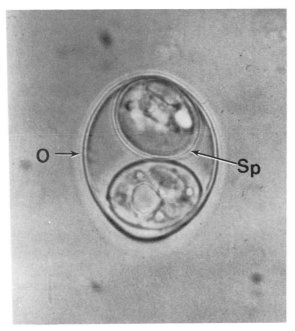

Figure 6.15 Photomicrograph of a fresh *Isospora felis* oocyst (O) containing two distinct sporocysts (Sp). The four sporozoites in each sporocyst are not in sharp focus and appear as an indistinct jumble of material.

in the bank vole, *Frenkelia microti*. It appears, therefore, that many if not all the eimeriine coccidia with an oocyst containing two sporocysts, each with four sporozoites, are heteroxenous (have more than one type of host in their life cycle). Some may still fit the criteria of the classically defined genus *Isospora*.

How the taxonomic problems arising from the elucidation of the life cycles of the isosporan coccidia will be resolved is at present not known. As we have already mentioned, some workers have adopted the drastic solution of suppressing all the other generic names applied to organisms not previously known to be isosporans, on the grounds of their being junior synonyms of *Isospora*. Others have retained the later names, as we have in this book, and allowed the name *Isospora* to fall out of use as the discovery of the complete life cycles of more and more of its erstwhile species allows them to be transferred to one of the other, "sarcocystic", genera. Perhaps a useable compromise would be to retain the generic name *Isospora* for all members of the group, differentiating them at the subgeneric level into *Sarcocystis*, *Toxoplasma*, etc.

Dogs and cats are commonly the final hosts of isosporan coccidia: *Toxoplasma* (cats only), *Sarcocystis* (cats, dogs, and wild felids and

141

canids), and *Besnoitia* (cats and wild felids). Other carnivorous or omnivorous mammals (including man) also serve as final hosts of some species of *Sarcocystis*, as do snakes and birds of prey (see below, and Tables 6.3 & 6.4.). It is probable that the evolutionary strategy of adopting infection of prey species with the enhanced possibility of infecting the predator, presumably the primary host, conferred considerable advantage on the isosporan coccidia.

Coccidiosis in domestic animals

Coccidiosis is an important disease of domestic animals (Tables 6.2, 6.3 & 6.4). Coccidia of the genus *Eimeria* and the various isosporan genera are the major causes of intestinal coccidiosis; *Cryptosporidium* is assuming prominence also (and may infect man). Coccidiosis of chickens and other fowl is of greatest economic importance, but coccidiosis is also common and severe in cattle, dogs, cats, and rabbits. Usually, severity of coccidiosis is directly related to the numbers of sporulated oocysts ingested by the susceptible host. This is because most coccidial infections are self-limiting, as asexual reproduction does not continue indefinitely. Therefore, the risk of clinical coccidiosis increases with the introduction of management practices that result in concentration of susceptible animals in limited spaces.

Cryptosporidium is a coccidian that is currently implicated as causing calf diarrhea. Oocysts have been described in feces of cattle and infection has been transmitted by administration of infected ilial scrapings by gavage. Severe infections have also occurred in immuno-compromised human patients, e.g. those suffering from acquired immune deficiency syndrome (AIDS).

Most species of coccidia infect epithelial cells of the intestinal tract. The coccidia infect certain zones of the bowel preferentially. In chickens, for example, *E. tenella* concentrates in the ceca and *E. necatrix* in the central section of the small intestine. *Eimeria truncata* grows in the kidney tubules of geese and *E. stiedai* infects the bile ducts of the livers of rabbits.

Damage to the epithelial lining of the bowel as a result of infection is a factor in the pathology of coccidiosis. Ingestion of large numbers of sporulated oocysts by a susceptible host may result in extensive damage to the epithelium lining the bowel. This damage, if very severe, may result in hemorrhage, fluid loss and interference with absorption of nutrients, and may open the path for secondary infection.

Immunity develops following infection. If exposure levels are not large, the disease which develops will be mild and the immunity resulting will provide the animals with the ability to withstand subsequent infections with the same species of coccidia. In general, *Eimeria* species are host specific.

Table 6.3 Some of the more important species of isosporan-type coccidia and *Cryptosporidium**.

Host	Species	Habitat	Pathogenicity
pigs	*Isospora suis*	small intestine	mild
dogs and cats	*I. bigemina* *I. felis*	small intestine small intestine	high moderate
man	*I. hominis* *I. belli*	intestines	slight
various vertebrates (definitive host cat)	*Toxoplasma gondii*	small intestine, cat; many tissues and cells of other hosts	slight
various vertebrates; rodents, horses, cattle (definitive host unknown)	*Besnoitia jellisoni* *B. besnoiti* (cattle) *B. bennetti* (horses)	many tissues and cells of hosts; intestine of definitive host	slight
various vertebrates; rodents (definitive hosts unknown)	*Hammondia* spp.	many tissues and cells of hosts	slight
various vertebrates; sheep (definitive hosts cat, dog)	*Sarcocystis tenella* (sheep, goat) *S. lindemanni* (man)	muscle; intestine of definitive host	slight
various vertebrates (cattle, pigs, rabbits, turkeys, etc.) and man if immunocompromised†	*Cryptosporidium muris*	intestines	slight

* Some of the names of isosporan coccidia listed here appear as synonyms of sarcocysts listed in Table 6.4. Isosporan taxonomy is in a state of flux as new information about the life cycles accumulates.

† *Cryptosporidium* from man has been named *C. garnhami*, but is probably not a separate species.

Table 6.4 Some of the more important species of *Sarcocystis* of man and domestic animals (from Levine & Tadros 1980).

Species	Intermediate host(s)	Final host(s)	Important synonym(s)
S. bertrami	horse, ass	dog	*I. rivolta, I. bigemina*
S. capricanis	goat	dog	
S. cruzi	ox	dog, wild canids	*S. fusiformis, I. rivolta, I. bigemina*
S. cuniculi	rabbit	cat	
S. cymruensis	rat	cat	
S. equicanis	horse	dog	⎫ ? junior synonyms of
S. fayeri	horse	dog	⎭ *S. bertrami*
S. fusiformis	water buffalo	cat	
S. gigantea	sheep	cat	*S. tenella*
S. gracilis	red deer	dog	
S. hemionilatrantis	mule deer	dog, coyote	
S. hirsuta	ox	cat, wild cat	*S. fusiformis*
S. hominis	ox	man, wild primates	*I. hominis*
S. horvathi	chicken	dog, ? cat	
S. leporum	cottontail rabbit	cat, racoon	
S. levinei	water buffalo	dog	
S. miescheriana	pig	dog, wild canids	*I. rivolta, I. bigemina*
S. moulei	goat	?	
S. muris	house mouse	cat, ferret	
S. porcifelis	pig	cat	*S. miescheriana*
S. rileyi	ducks (domestic and wild)	? opossum	
S. suihominis	pig	man, wild primates	*I. hominis*
S. tenella	sheep	dog, wild canids	*I. rivolta, I. bigemina*

Treatment and control   Treatment of animals with clinical coccidio-sis, if undertaken, should be primarily supportive. Fluids to correct dehydration and maintain salt balance, and antibiotics to combat secondary infection are probably useful. Coccidiostats (drugs which control but do not eliminate the infection) may be given, but there is little evidence that they affect the course of the infection once clinical disease occurs.

As mentioned earlier, coccidiosis usually becomes a problem as a result of management practices which result in concentration of susceptible animals in limited spaces. Control of coccidiosis thus requires action to reduce ingestion of sporulated oocysts and to ameliorate the consequences of their ingestion. Ingestion of sporulated oocysts can be reduced by improvements in sanitation and by periodic movement of animals to clean areas. Coccidiostatic drugs are com-monly incorporated in the food and water of poultry and other domestic animals to prevent development of disease following infection. Immunization of poultry is practiced by feeding sporulated oocysts and then giving coccidiostats.

GENUS *TOXOPLASMA* NICOLLE
AND MANCEAUX, 1909

Elucidation of the complete life cycle of *Toxoplasma*, involving the role of cats in its transmission, and the consequent clarification of its taxonomic position, resulted from relatively recent work by Hutchi-son, Frenkel, and their colleagues, who first showed in 1965 that *T. gondii* could be transmitted in the feces of infected cats. They subsequently demonstrated that the transmissive stage in cat feces is an oocyst characteristic of the coccidia in the genus *Isospora*, thus settling the major taxonomic issues and resolving many questions about the epidemiology of *Toxoplasma* also. Some authors now treat *Toxoplasma* as a synonym of *Isospora*.

The forms of *Toxoplasma gondii* inhabiting cells other than gut epithelium were first described from a wild rodent (*Ctenodactylus gundi*) in North Africa half a century earlier, in 1908. The parasite can be freely transmitted (by experimental inoculation) from one host species to another; there is no morphological difference between the organisms found in different host species.

Toxoplasma gondii can probably infect all warm-blooded animals (mammals and birds), but not cold-blooded ones. *Toxoplasma* has been recorded from all parts of the world. It may cause acute illness (toxoplasmosis) and death of infected persons (or other animals), but usually infection is completely inapparent. It is likely that about one in

145

every four persons reading this book has been infected with *T. gondii* at some time without having been aware of it. While *T. gondii* is capable of reproducing asexually in a wide variety of hosts, merogony and gametogony are restricted to the domestic cat (*Felis domestica*) and perhaps a few of its close wild relatives.

Morphology and life cycle

The life cycle of *Toxoplasma* (Fig. 6.16) in the primary or definitive host, the cat, is similar to that of other Eimeriina (Fig. 6.10). The stages in cells of the feline intestinal epithelium include merogony with the production of merozoites (Fig. 6.17) and gametogony with the production of gametocytes. The gametocytes mature to yield gametes which fuse to produce a zygote. The zygote develops a protective wall and is converted into an oocyst. The oocyst is passed from the cat in its feces and, after undergoing sporulation, is infectious for probably any

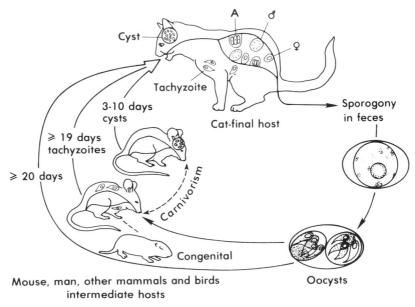

Figure 6.16 Life cycle of *Toxoplasma*. Cats, the definitive hosts, can become infected by ingesting infected animals or sporulated oocysts. Prepatent periods to the shedding of oocysts vary with the stage of *T. gondii* ingested: 3–10 days after ingesting tissue cysts, 19 days or longer after ingesting tachyzoites or oocysts. In nature it is the tissue cyst-induced cycle in cats that is important because only one-half or fewer of previously uninfected cats which ingest tachyzoites or oocysts subsequently shed oocysts, whereas almost every previously uninfected cat sheds oocysts after ingesting tissue cysts. Asexual forms (A) of *Toxoplasma*, and gametocytes (♂ male, and ♀, female) both occur in the feline host (from J. P. Dubey 1976).

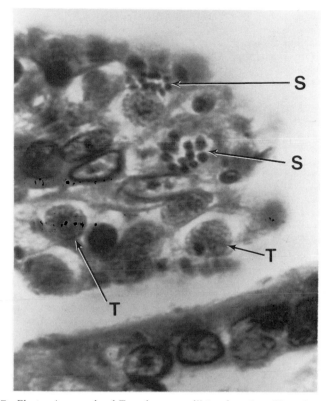

Figure 6.17 Photomicrograph of *Toxoplasma gondii* trophozoites (T) and meronts (S) in intestinal mucosa of a cat.

other warm-blooded animal that should ingest or inhale it. The ripe oocyst contains two sporocysts, each containing four sporozoites.

In hosts other than the definitive host, *Toxoplasma* does not complete its sexual development but, like many parasites which enter the "wrong" host, wanders through the body in an almost aimless way. In these secondary hosts, only one stage is known: a small crescentic organism called a zoite (sometimes "trophozoite"), slightly pointed at one end with a central nucleus (Fig. 6.18), The zoite is about 5 μm long and 1–2 μm broad and it can move with a gliding motion. Accounts of its reproduction vary, but it appears to divide asexually by a process called endodyogeny, or internal budding, in which two "daughters" develop within the parent. Electron microscopy reveals that the zoites possess a full complement of apicomplexan organelles: conoid, rhoptries, and micronemes (see Ch. 2).

In the early stage of infection, the zoites enter macrophages (both actively and by ingestion). They inhibit fusion of the phagosomes with

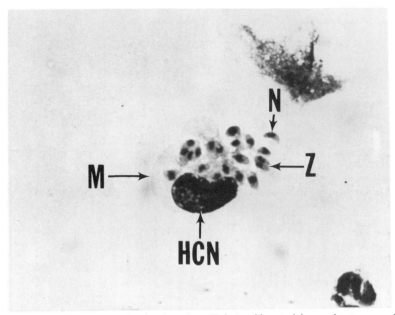

Figure 6.18 Photomicrograph of tachyzoites (Z) being liberated from a host macrophage cell (pseudocyst) in a smear of mouse peritoneal fluid. The tachyzoite nucleus (N) is centrally located in the banana-shaped organism. The host cell nucleus (HCN) and the host cell membrane (M), which in part still surrounds the tachyzoites, are visible (photomicrograph provided by J. P. Dubey).

lysosomes, and thus survive and divide until the cell is full of them. These aggregations of parasites, bounded only by the plasmalemma of the host cell, are called pseudocysts (Fig. 6.18), a true cyst being defined by parasitologists as one in which the protective membrane surrounding the parasite is produced at least in part by the parasite. Zoites within pseudocysts divide rapidly and hence are called tachyzoites (Greek *tachos*, "speed"). The host cell finally dies and bursts, liberating the zoites which re-enter other cells and continue the process. This "proliferative phase" of the infection occurs in all the viscera and in the circulating blood. The development of immunity seldom results in complete destruction of the parasites.

When the proliferative stage ends, some of the zoites will be found in true cysts with thin, tough walls. These zoites are called bradyzoites as they develop slowly (Greek *brady*, "slow"). Tissue cysts (Fig. 6.19) are found throughout the body, especially in the central nervous system, musculature, and lungs. When they are fully grown (up to 60 μm in diameter), all trace of the host cell may be lost. Chronic infections with the parasites present only in tissue cysts may last years. Such latent infections may become active again in individuals who are

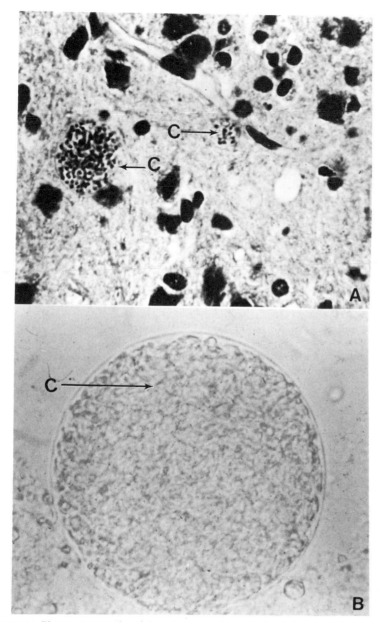

Figure 6.19 Photomicrographs of *Toxoplasma gondii* cysts. In (a) the cysts (C) are present in a section of brain tissue. In (b) a single cyst (C) that has been released from a brain by homogenization can be seen at higher magnification.

artificially immunosuppressed during cancer therapy or following organ transplants, or naturally if infection with the virus which causes acquired immunodeficiency syndrome (AIDS).

Transmission and epidemiology

Toxoplasma gondii can be transmitted by various contaminative methods including ingestion of mature oocysts from cat feces. The common presence of cats in houses and barns makes this a likely mode of transmission to man and to herbivorous animals. The discovery of cysts in the walls of, and lying free in, lung alveoli of infected experimental animals suggests that inhalation is also a mode of infection. The parasite is common in many domestic animals, such as cattle, sheep, and dogs. From cattle and sheep it may easily be transmitted to man and other meat eaters by the eating of raw infected meat (however, even light cooking kills the organism).

Infection in man and other animals also occurs by the congenital route. If a pregnant female has an acute infection, the chance of the parasites passing through the placenta and infecting the embryo is fairly high.

Pathogenesis

In man and, as far as is known, in other animals, toxoplasmosis which has been acquired after birth may result in fever and swelling of the lymph glands, or may be so mild as to pass unnoticed. Generalized infection, in which brain, lungs, liver, and other tissues are involved, may occur and may cause death. This is the acute phase of toxoplasmosis. If death does not ensue, the acute infection proceeds to the chronic latent phase in which only encysted forms are present and no sign of disease is seen.

Since there is serological evidence that about one-quarter to one-third of the population of England and the USA has at some time been infected with *T. gondii*, yet fewer than two hundred clinical cases occur annually, it is obvious that the development of severe or fatal illness is a very rare consequence. Infection acquired before birth, though quite rare (probably less than one per one thousand live births), usually causes damage. By the time the infected child is born, infection may have devastated large areas of the brain. The retina of the eye is also infected and this may lead to blindness. In relatively mild congenital toxoplasmosis, damage to the retina may be the only detectable sign. Severely infected infants are often either still-born or die soon after birth.

150

Diagnosis

A clinical diagnosis of suspected toxoplasmosis is confirmed by the isolation of the parasite from material such as lymph gland or tonsil tissue by inoculation into mice. There are also various serological procedures which detect the presence of antibodies to *T. gondii*; a positive serological test indicates experience of the parasite at some time. If two serological tests are done at intervals of a few weeks, and the second shows a higher titer (i.e. a more strongly positive reaction) than the first, then acute toxoplasmosis is possible. The most commonly used tests include a complement-fixation reaction using an extract of *T. gondii* zoites as antigen, an agglutination test (in which killed zoites are caused to adhere to one another when treated with serum containing antibody), an indirect fluorescent antibody test, and the so-called "dye test" which depends on the fact that specific antiserum affects living zoites in such a way that they do not become stained when immersed in methylene blue solution, whereas normal living zoites are stained. Such tests have given positive results in up to 77% of cats, 22% of dogs, 50% of pigs, 64% of sheep, 21% of cattle, and 25–30% of human beings in England and the United States of America, while higher figures have been obtained for man in other regions, e.g. 94% in Guatemala.

Treatment and control

Severe congenital toxoplasmosis has usually done its damage before birth, so treatment is of little use. Post-natal infection, however, can be treated by a mixture of the antimalarial drug pyrimethamine and a sulphonamide. If the patient is pregnant, the antibiotic spiramycin should be used instead. As noted earlier, pre-natal infection usually occurs only if the woman becomes acutely infected while pregnant. Immune carriers seldom, if ever, produce pre-natally infected infants.

Ideally, a woman who is planning to bear a child in the near future should be serologically tested for anti-*Toxoplasma* antibodies. If the test is positive with a fairly low, not rising, titer, she is almost certainly immune and therefore her fetus will not be at risk. If the test is negative, however, the woman should take reasonable precautions to avoid becoming infected while pregnant. Such precautions could include avoiding undue familiarity with cats, careful hand-washing after unavoidable contact with them, not emptying the family cat's litter tray; she should also avoid handling or eating raw or under-cooked meat. More advanced precautions, such as placing the domestic pet into a cat boarding kennel during the pregnancy, are probably unnecessary since congenital infection with toxoplasmosis is, fortunately, rare.

GENUS *SARCOCYSTIS* LANKESTER, 1882

Many species of this genus have been described. It is very common (70–100%) in some herbivorous mammals (cattle, sheep, horses), and has also been recorded in pigs, monkeys, rabbits, rodents, ducks, chickens, man (rarely), and many other species throughout the world. Almost always it seems entirely nonpathogenic, though fatal infections have been recorded in mice, and in man it may cause mild disease. The discovery of typical eimeriine merogony and gametogony in a variety of carnivorous definitive hosts has greatly changed and somewhat confused the nomenclature of *Sarcocystis*, which, like *Toxoplasma*, some authors now synonymize with *Isospora*. Species of *Sarcocystis* which parasitize man or domestic animals, either as intermediate or final hosts, are listed in Table 6.4. The stages in the intermediate hosts are the long-known muscle "sarcocysts", while those in the final hosts are the isosporan intestinal stages undergoing gametogony and sporogony. Some of the latter have also been known for a long time, but their participation in the life cycle of *Sarcocystis* has been recognized only in the last decade or two. Many final hosts are as yet unidentified. All must be carnivores, or at least omnivores. Most of those known are mammals, but some are predatory birds or snakes. Examples include *S. cernae* (vole and kestrel), *S. dispersa* (house mouse and barn, masked or long-eared owls), *S. scotti* (house mouse and tawny owl), *S. sebekei* (field or house mice and tawny owl), *S. idahoensis* (deer mouse and gopher snake), *S. lampropeltis* (unknown intermediate and king snake), *S. roudabushi* (unknown intermediate and gopher snake), and *S. singaporensis* (rat and reticulated python).

The zoites of all species are almost entirely restricted to the muscle fibers (including cardiac muscle) of their hosts (rarely have they been recorded from brain). Here they are seen as large, sometimes very large, oval or elongated cysts, which may be as long as 1–2 mm (Fig. 6.20). The larger cysts are divided into irregular compartments by a network of cytoplasmic partitions called "trabeculae", and their central regions may be more or less empty. The cyst is lined by a layer of protozoan cytoplasm which contains many nuclei, not separated from one another by cell walls. It is from this lining layer that the trabeculae develop. The wall itself is complex. On the outside is a layer of host connective tissue, and within this is a spongy, or fibrous, layer (or layers) of uncertain origin. From the layer of nucleated cytoplasm lining the cyst wall, rounded cells are budded which divide and give rise to the zoites which divide by endodyogeny. The zoites (Fig. 6.21) closely resemble those of *Toxoplasma* and *Besnoitia* in their general shape and structure, but they are larger (10–15 μm long).

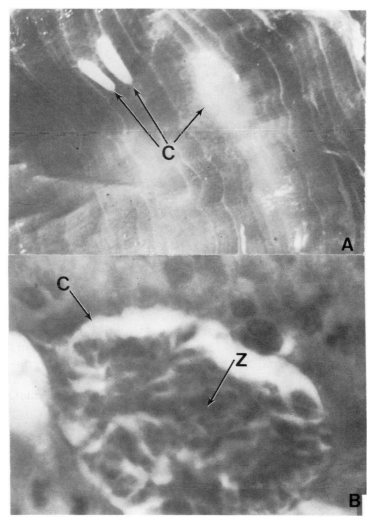

Figure 6.20 Photomicrographs of sarcocysts in muscle. In (a), a low magnification photomicrograph, the cysts (C) appear as oval, white structures in the muscle. In (b), a higher magnification photomicrograph, the mass of zoites (Z) can be seen in the cyst (C).

Sarcocystis tenella infects sheep in many, if not all, countries in the world. The sexual phase occurs in dogs and wild canids. In sheep, the cyst is large enough to be seen with the naked eye. Although apparently nonpathogenic, the cyst contains a very powerful toxin ("sarcocystin") which, if extracted and injected to rabbits, is lethal in doses as low as 0.05 mg kg^{-1} body weight. Given orally, however, the toxin is harmless (which is perhaps fortunate since most nonvege-

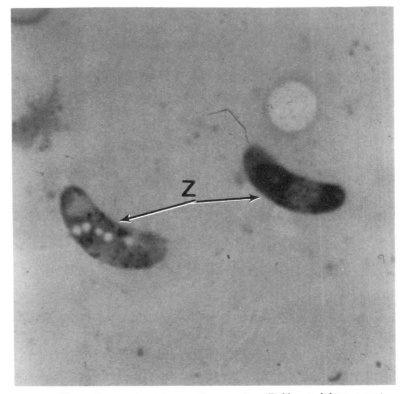

Figure 6.21 Photomicrograph of *Sarcocystis* sp. zoites (Z) liberated from a cyst.

tarians must consume large numbers of cysts of *Sarcocystis* in mutton and beef). Other species have cattle, horses, pigs, ducks, and many wild animals as intermediate hosts.

Sarcocystis lindemanni is the name originally given to the species found encysted in muscle in man. This name may well include accidental infections of man with several species normally infecting other primates and (in their sexual, "isosporan", phase) various carnivores. *Sarcocystis lindemanni* has been seen only very rarely in human muscle as the cysts are too small to detect with the naked eye. However, if normally nonpathogenic, sarcocyst infection of man may not be quite so rare as this would indicate.

If the infection is heavy, degeneration of the surrounding muscle fibers and consequent muscle weakness results, with some pain. Man is now known to be the definitive host of other species of *Sarcocystis*, development of which, in the intestinal wall, can cause quite severe diarrhea; examples are *S. bovihominis* (intermediate host, oxen) and *S. suihominis* (intermediate host, pig).

GENUS *BESNOITIA* HENRY, 1913

This genus is also sometimes treated as a synonym of *Isospora*. Species of this genus infect cattle, horses, deer, rodents or lizards, but not man. All known species are seen as zoites, like those of *T. gondii* but a little larger (5–9 × 2–4 μm), and inhabit large (100–500 μm) cysts (Fig. 6.22). These cysts have a thick collagenous wall containing numerous host nuclei, probably belonging to an original single host cell which has become greatly enlarged and multinucleate. Within the cyst the zoites multiply by binary fission or, possibly, endodyogeny. *B. jellisoni* will infect (and may kill) laboratory mice, and the cysts occur in the mesenteries.

Besnoitia besnoiti of cattle (in Europe and Africa) and *B. bennetti* of horses (Europe, Africa, and North America) inhabit the skin (and the cornea). The definitive hosts of *B. besnoiti* and some other species are wild or domestic felids; those of other species are, as yet, unknown. *Besnoitia* is of some economic importance since, although the disease which it produces is chronic, infected animals lose condition and the hair drops out of infected areas of skin. The mortality rate is about 10% and no reliable treatment is known.

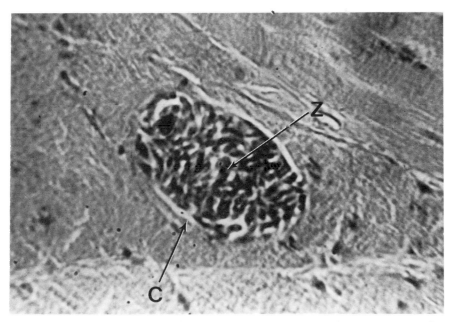

Figure 6.22 Photomicrograph of a small cyst (C) of *Besnoitia* sp. in the tongue of a reindeer. The cyst contains many zoites (Z).

ADDENDUM

Pneumocystis carinii Delanoë and Delanoë, 1912, is a parasite of uncertain taxonomic position. Opinions differ as to whether it is a protozoon or a fungus. It has even been identified as a pathologically modified mitochondrion of its host's pneumocytes. Whatever it may be, it causes a disease of man known as atypical interstitial plasma-cell pneumonia, and has been recorded from man, dogs, and rodents in North and South America, Europe, Australia, and China.

The organisms (if such they are) lie free in the alveoi of the lung. They are spherical, 7–10 μm in diameter, and consist of an outer capsule (apparently polysaccharide) and an inner uninucleate proto-plasmic body about 1–3 × 1 μm in size. The whole organism (capsule

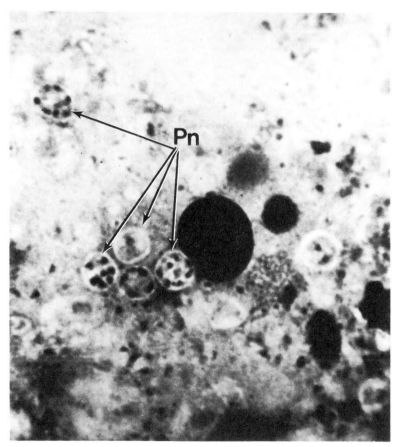

Figure 6.23 Photomicrograph of *Pneumocystis carinii* (Pn) in an impression smear from an infected rat lung.

and protoplasm) divides by binary fission. Another stage (Fig. 6.23) is known, in which the capsule is larger (10–12 μm). The inner body is also enlarged and divides by an apparent meiosis into eight uninucleate organisms (about 1.5 × 1 μm) within the capsule (or "cyst"). This stage is presumed to be infective. The uninucleate organisms presumably fuse in pairs after emergence, to restore the diploid condition. The detailed morphological descriptions summarized here in no way support the supposed "mitochondrial" nature of *P. carinii*.

Pneumocystis carinii is relatively most common in infants (especially premature ones), particularly in central Europe where epidemics have been recorded in maternity and children's hospitals. *Pneumocystis carinii* infection has also been seen in adults who have been receiving prolonged treatment with corticosteroids, which is known to reduce the capacity to synthesize antibodies. The total number of human infections known is probably not more than a few thousand, although with increased use of treatments which result in immunosuppression in organ transplantation and cancer therapy, infections are becoming more common. The disease has also been reported from patients suffering from acquired immune deficiency syndrome (AIDS). The parasite has been found only rarely in animals other than man and the rat. It is likely that it may be more common than has been thought, but becomes pathogenic only in immunocompromised hosts. Probably either chronically infected adults or domestic animals (or both) may serve as reservoir hosts, transmission occurring by droplet infection. Transplacental infection has also been proposed.

The pneumonia results from the fact that the parasites, together with the host's plasma cells, block the lung alveoli and bronchioles; also, the alveolar walls become thickened and infiltrated with plasma cells. Characteristically no fever is produced. Apart from the clinical picture of an afebrile pneumonia which does not respond to treatment with antibiotics, confirmatory diagnosis is difficult and may depend on finding the organisms in smears prepared from a piece of the lung removed at biopsy (or necropsy). The death rate in untreated patients is high (about 80%).

For treatment, the trypanosomicidal compound pentamidine isethionate may be beneficial. A less toxic treatment, more recently introduced, involves the use of a sulphonamide–diaminopyrimidine mixture such as co-trimoxazole (sulphamethoxazole plus trimethoprim). Both constituents are folic acid inhibitors, though active in different ways.

FURTHER READING

Beautyman, W. 1983. *Pneumocystis carinii* is an endogenous liposomally modified mitochondrion. *Medical Hypotheses* **10**, 281–9.

Dubey, J. P. 1977. *Toxoplasma, Hammondia, Besnoitia, Sarcocystis* and other tissue cyst-forming coccidia of man and animals. In *Parasitic protozoa*, Vol. 3, J. P. Kreier (ed.), 101–237. New York: Academic Press.

Fox, J. E. 1978. Bovine coccidiosis. *Modern Veterinary Practice* **59** 599–603.

Levine, N. D. 1973. *Protozoan parasites of domestic animals and man*, 2nd edn. Minneapolis, Minnesota: Burgess.

Levine, N. D. 1982. Apicomplexa. In *Synopsis and classification of living organisms*. New York: McGraw-Hill.

Levine, N. D. 1985. *Veterinary protozoology*. Ames, Iowa: Iowa State University Press.

Levine, N. D. & W. Tadros 1980. Named species and hosts of *Sarcocystis* (Protozoa: Apicomplexa: Sarcocystidae). *Systematic Parasitology* **2**, 41–59.

Long, P. L. 1982. *The biology of the Coccidia*. Baltimore, Maryland: University Park Press.

Matsumoto, Y. & Y. Yoshida 1986. Advances in *Pneumocystis* biology. *Parasitology Today* **2**, 137–42.

Noble, E. R. & C. A. Noble 1982. *Parasitology*. Philadelphia, Pennsylvania: Lea and Febiger.

Pohlenz, J., H. W. Moon, N. F. Cheville & W. J. Bemrick 1978. Cryptosporidiosis as a probable factor in neonatal diarrhea of calves. *Journal of the American Veterinary Medical Association* **172**, 452–7.

Soulsby, E. J. L. 1982. *Helminths, arthropods and protozoa of domesticated animals*, 7th edn. London: Baillière Tindall.

Tadros, W. & J. J. Laarman 1982. Current concepts on the biology, evolution and taxonomy of tissue cyst-forming eimeriid Coccidia. *Advances in Parasitology* **20**, 293–468.

CHAPTER SEVEN

Malaria parasites and their relatives

The term "malaria parasites" is generally restricted to species of the genus *Plasmodium* which inhabit reptiles, birds and mammals. Use of the word "parasite" in this phrase is so well established that it would require *force majeure* to change it to "symbionts", and indeed such a change is probably not necessary as, although by no means all malaria parasites and their relatives are significantly harmful to their hosts, their intracellular habit leads inevitably to destruction of host cells and thus they must be regarded as at least potential pathogens. *Plasmodium* is classified in the family Plasmodiidae of the phylum Apicomplexa; other closely related genera which do not infect man are placed in the families Haemoproteidae and Leucocytozoidae. All of these genera are members of the suborder Haemosporina. They are obligate intracellular parasites (or symbionts) for almost all of their life cycles and have two hosts: a vertebrate in which asexual reproduction, or merogony, occurs and an invertebrate (always a blood-sucking dipterous insect) in which fertilization occurs, the insect host being regarded as the vector.

Merogony is a type of asexual reproduction possibly restricted to the Apicomplexa, resulting in the roughly simultaneous production of a number of progeny (merozoites). In this process all, or almost all, the nuclear divisions are completed before the progeny begin to bud from the parent cell; it is sometimes called "schizogony."

FAMILY PLASMODIIDAE

These organisms are placed in the single genus *Plasmodium* Marchiafava and Celli, 1885, a genus of Apicomplexa in which fertilization occurs in mosquitoes (Diptera, Culicidae) and merogonic asexual reproduction occurs in vertebrates. In the latter host, merogony takes place in fixed tissue cells and in erythrocytes, and gametogony takes place in erythrocytes. The intraerythrocytic forms of *Plasmodium* metabolize hemoglobin, producing a characteristic yellow, brown or black "malarial pigment" (hemozoin), a compound containing heme, in vacuoles within their cytoplasm. It has been suggested that the genus *Plasmodium* should be split into three or more genera, but this division, if accepted at all, is now usually restricted to the subgeneric level. The following subgenera may be recognized.

(a) In primates: *Plasmodium* and *Laverania* (transmitted by mosquitoes of the genus *Anopheles* only).
(b) In lemurs and lower mammals: *Vinckeia* (transmitted only by *Anopheles*).
(c) In birds: *Haemamoeba*, *Huffia*, *Giovannolaia* and *Novyella* (transmitted by various genera of mosquitoes, anopheline and culicine).
(d) In reptiles: *Sauramoeba*, *Carinamoeba* and *Ophidiella*. The vectors of most species are unknown, but *P. (Sauramoeba) mexicanum* is transmitted by Phlebotomimae (sandflies) and *P. (S.) agamae* develops in *Culicoides* (midges).

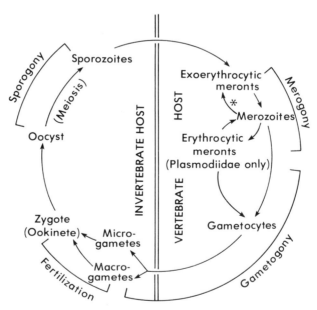

* Not in *Plasmodium* spp. of mammals

Figure 7.1 Diagrammatic representation of the life cycle of members of the suborder Haemosporina.

The life cycles of all species of *Plasmodium* are very similar (Fig. 7.1) and, indeed, have many similarities with those of the suborder Eimeriina (see Fig. 6.10). The vertebrate host is infected by means of small (10–15 ×0.5–1 μm), fusiform, uninucleate sporozoites (Fig. 7.2)

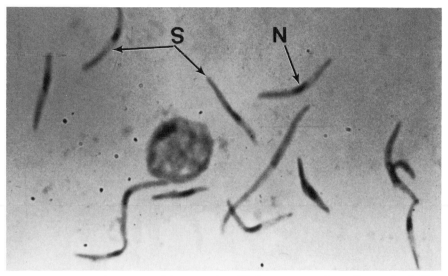

Figure 7.2 *Plasmodium cynomolgi* sporozoites (S) from a mosquito salivary gland. The sporozoite nucleus (N) is in the center of the elongate organism.

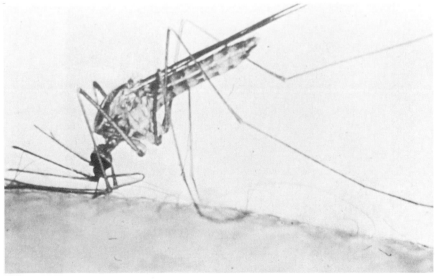

Figure 7.3 Anopheline mosquito feeding. It is during the process of feeding that the gametocytes are ingested and the sporozoites are introduced into the vertebrate host.

161

injected with the saliva of an infected mosquito when the latter feeds (Fig. 7.3). Only female mosquitoes feed on blood, so only they can act as vectors. The sporozoites penetrate fixed tissue cells and commence merogony. The cells in which this occurs differ in the different groups of subgenera. In species parasitizing mammals this first cycle of merogony (exoerythrocytic merogony) occurs only in liver parenchyma cells (Fig. 7.4), though there is evidence that sporozoites may reach these cells via Kupffer cells. In the subgenera which infect birds and reptiles the primary exoerythrocytic meronts are usually in vascular endothelium or macrophage cells at the site of the bite. Some of the merozoites resulting from the primary exoerythrocytic meronts invade other vascular endothelial or macrophage cells throughout the body to produce secondary exoerythrocytic meronts (Fig. 7.5c). Some of the merozoites produced by these secondary meronts continue to infect vascular endothelial and macrophage cells, while others enter red blood cells (or their precursors) to initiate erythrocytic merogony, both

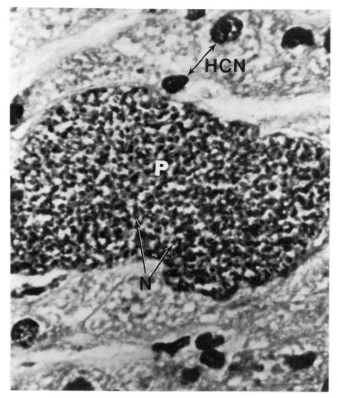

Figure 7.4 *Plasmodium vivax* exoerythrocytic meront. The large parasite (P) contains many nuclei (N). Host cell nuclei (HCN) are present in the surrounding liver tissue.

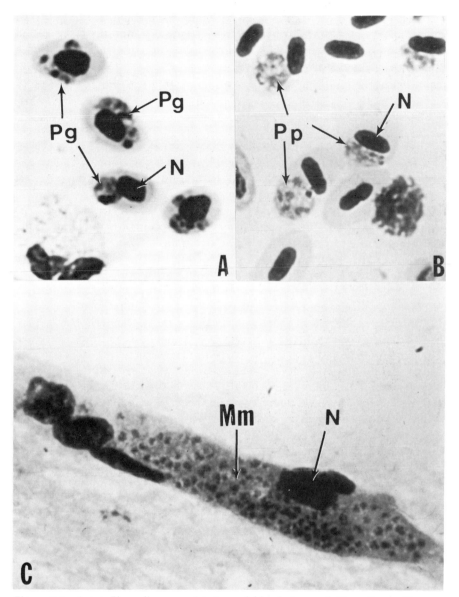

Figure 7.5 Avian *Plasmodium* spp. (a) Asexual blood stages of *Plasmodium gallinaceum* (Pg). (b) Asexual blood stages of *Plasmodium praecox* (Pp); the intensely stained structures are erythrocyte nuclei (N). (c) Exoerythrocytic meront (Mm) of *P. gallinaceum* in vascular endothelium; nuclei of vascular endothelial cells (N) can be seen in this crush preparation of an infected tissue.

types of merogony continuing side by side. The terminology of the exoerythrocytic stages in birds is rather confused: the first generation exoerythrocytic meronts are called primary exoerythrocytic (or pre-erythrocytic or cryptozoic) meronts; subsequent generations are called secondary exoerythrocytic (or exoerythrocytic, metacryptozoic, or phanerozoic) meronts (the terms in parentheses are less desirable).

In 1948, H. E. Shortt and P. C. C. Garnham published the first description of primary exoerythrocytic meronts ("pre-erythrocytic schizonts") of a malaria parasite of primates (*P. cynomolgi*). Similar forms in other species, including those parasitizing man, were described by them and their collaborators during the ensuing decade, and the hypothesis of repeated exoerythrocytic merogonic cycles in the liver was developed. This seemed to account for the occurrence of relapses of malaria after months or years during which parasites were absent from the blood. However, further work by Garnham and his followers indicated that repeated cycles of exoerythrocytic merogony do not occur in malaria parasites of mammals. True relapses or recurrences (i.e. reappearance of parasites in the blood after periods of complete absence therefrom) are now thought to occur in only two of the species of *Plasmodium* which normally infect man (*P. vivax* and *P. ovale*), in a few closely related species which infect monkeys (including *P. cynomolgi*), and in at least one species infecting rodents (*P. yoelii*). Such recurrences, and the prolonged prepatent periods between injection of sporozoites and the appearance of parasites in red blood cells which are characteristic of some strains of *P. vivax* from temperate regions, are now believed to be due to the presence in liver cells of dormant stages which are reactivated by some as yet unknown endogenous or exogenous stimulus. In *P. vivax* and *P. cynomolgi* these dormant stages have been identified as a special type of sporozoite, the "hypnozoite", morphologically indistinguishable from a "normal" sporozoite, which occurs in different proportions in different strains. Hypnozoites probably occur in *P. ovale* also. In the relapsing parasite of rodents, *P. yoelii*, dormant or retarded meronts (with 30–300 nuclei) have been discovered in the liver many months after natural or experimental infection, and presumably relapses result from reactivation or continued slow development of these forms. In other species, apparent "relapses" result only from recrudescences of small populations of intraerythrocytic parasites which have survived the host's defensive mechanisms by some means not yet fully understood – perhaps by changing their surface antigens, or by hiding from antibodies in small secluded visceral blood vessels until the host's antibody titer drops, or by a combination of both methods.

In the erythrocytes the merozoites of all species of *Plasmodium* develop similarly. At first they are seen as small ring-shaped parasites; they then grow (during which process they are known as tropho-zoites); and finally they commence nuclear division. Some of the merozoites in erythrocytes in both birds and mammals develop, not as erythrocytic meronts, but as sexual individuals or gametocytes. The morphology of the blood stages of *Plasmodium* species infecting humans is shown in Fig. 7.6.

One of the unsolved problems in the Eucoccidiida as a whole is the elucidation of the factors determining whether a merozoite behaves as a male, female or asexual individual. The male and female gametocytes (which can be differentiated in stained blood films as the latter have a more compact nucleus and very basophilic cytoplasm, containing many ribosomes) do not divide, but remain within their host erythrocytes until they either die or are ingested by a mosquito in which they continue their development in the insect's stomach. In the mosquito the female emerges from the red cells, rounds up (if not already spherical) and, now a mature gamete, awaits the arrival of a male. Meanwhile, the male gametocyte becomes very active: emerging from its host cell, having undergone three nuclear divisions it develops eight flagella, all in the space of about 15–20 min; the flagella emerge from the surface of the gametocyte (this process is termed "exflagellation": Fig. 7.7a); one nucleus passes inside the membrane surrounding the axoneme of each flagellum, and the resulting male gametes break off the gametocyte and swim rapidly away: those which meet female gametes, penetrate and fertilize them (Fig. 7.7b). The process of gamete-formation, of both sexes, is called gametogony. As in the Eimeriidae, the females are termed macrogametocytes and macro-gametes, while the males are often spoken of as microgametocytes and microgametes. The fertilized female elongates into a motile ookinete (Fig. 7.7c). The male and female nuclei fuse at this stage or shortly after. The ookinete burrows into and, usually, through a cell of the single-layered epithelium forming the wall of the mosquito's stomach. On the outer surface of the stomach wall it encysts and becomes an oocyst (Fig. 7.7d). The oocyst contents then undergo sporogony, a process similar to merogony, which results in the production of thousands of uninucleate sporozoites. The first nuclear division of sporogony is a reduction division and all stages except the ookinete are haploid. The oocyst, which has been increasing in size throughout sporogony, bursts and the liberated sporozoites spread throughout the mosquito's hemocoel. Most of them eventually penetrate the salivary glands, where they remain until the insect has its next blood meal and they are injected with the saliva into the animal on which it is feeding. If the latter is a susceptible host, they will then enter the appropriate

165

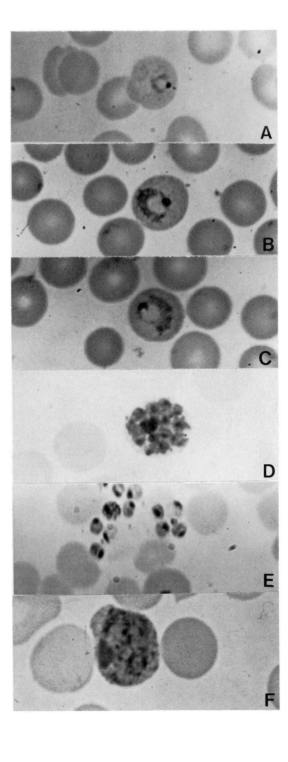

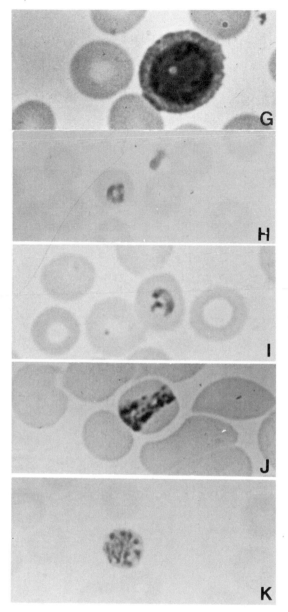

Figure 7.6 The morphology of the species of *Plasmodium* which infect man. (a)–(g) *Plasmodium vivax*: (a) young trophozoite (ring form) in human blood cell; (b) & (c) older trophozoites in erythrocytes which are enlarged and contain Schüffner's dots; (d) mature meront; the individual merozoites are already formed but still retained in the parasitophorous vacuole; (e) merozoites released from broken meront; (f) female and (g) male gametocytes.

(h)–(l) *Plasmodium malariae*: (h) young trophozoite ("ring"); (i) older trophozoite; (j) band form; (k) early meront; (l) mature meront.

continued

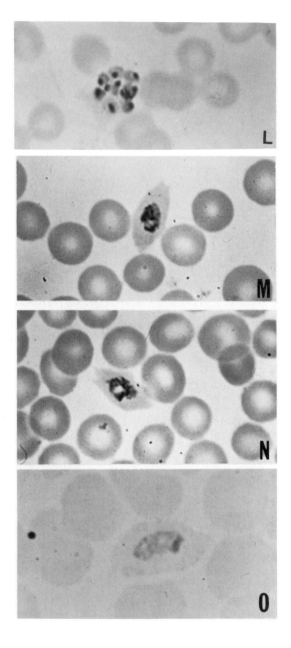

(m)–(o) *Plasmodium ovale*: (m) & (n) immature meronts in distorted erythrocytes; (o) trophozoite in an enlarged and distorted erythrocyte, with Schüffner's dots just visible.

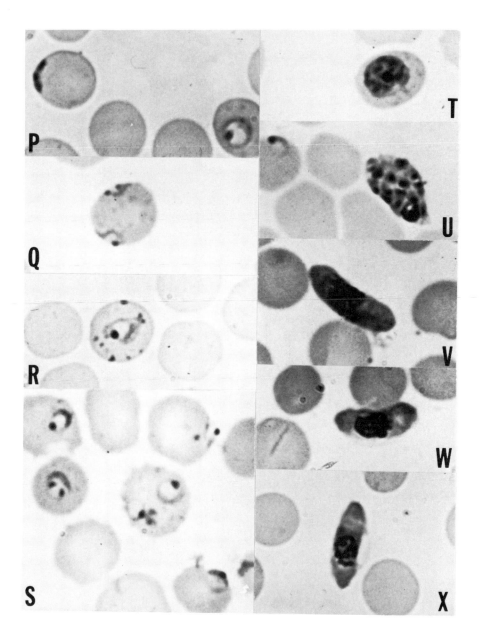

(p)–(x) *Plasmodium falciparum*: (p) trophozoites in erythrocytes, accolé form (left) and ring form (right); (q) two accolé forms in one erythrocyte, these are more advanced than those shown in (p); (r) ring-form trophozoite in an erythrocyte containing Maurer's clefts; (s) a heavily infected group of erythrocytes, some containing one and some more than one trophozoites; (t) a young meront; (u) a mature meront; (v) male, and (w) & (x) female gametocytes.

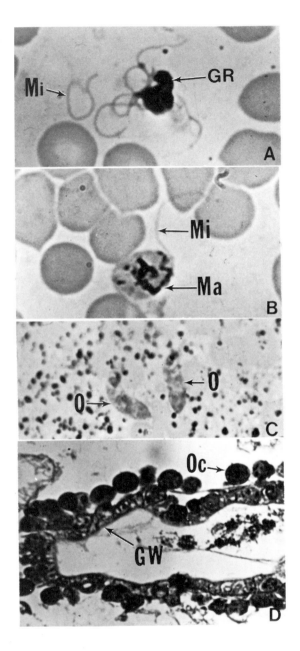

Figure 7.7 Stages of *Plasmodium* in the vector. (a) Development of male microgametes (Mi) from microgametocyte ("exflagellation"); after exflagellation a gametocyte residue (GR) remains (slide made from preparation *in vitro*). (b) Fertilization: the microgamete (Mi) enters the macrogamete (Ma). (c) Ookinetes (O) develop in the mid-gut of the mosquito. (d) Oocysts (Oc) develop from the ookinete on the outside of the gut wall (GW) of the mid-gut of the mosquito.

tissue cells and commence exoerythrocytic merogony. This complicated life cycle is summarized diagrammatically in Figure 7.1. The development in the mosquito appears to have no harmful effect on the insect.

Malaria in man

Four species of *Plasmodium* have man as their main (or only) vertebrate host. Their geographical distribution, and main distinguishing characteristics, are summarized in Table 7.1. All are transmitted by female mosquitoes of the genus *Anopheles*.

Plasmodium (Plasmodium vivax) (Figure 7.6a–g)   This species is the most widespread of the human malaria parasites, and used to be common in southern England. (Oliver Cromwell is said to have suffered from it.) It existed precariously in England, where winters are too cold for adult mosquitoes and where transmission could occur only in the summer; for the rest of the year it survived in its vertebrate hosts. Some strains of *P. vivax* from temperate regions adapted to this situation by developing very long periods of prepatency (up to nine months) before invading red blood cells, during which time they persist as "dormant" hypnozoites ("sleeping forms") within liver cells. They thus survive from one season of vector availability to the next, without being exposed to the risk of eradication by the host's defenses. Increasing urbanization and draining of swamps, etc., which reduced the contact between *Anopheles* and man, probably led to the extinction of malaria in England, though a few indigenous infections were recorded after the 1914–18 war, the mosquitoes deriving their infections from returning soldiers. Subsequently, malaria eradication programs have removed malaria from many of its temperate habitats.

   Plasmodium vivax produces a relatively mild disease, benign tertian malaria, in man, which, though debilitating, is seldom fatal. If untreated, it persists for many years. The adjective "tertian" is derived from the fact that the characteristic periodic fevers, which are associated with the rupture of erythrocytic meronts, usually occur at intervals of 48 hours: thus, if the day of the first fever is numbered 1 (after the old Roman fashion), the next fever will occur on day three. The adjective "benign", which would probably be disputed by those who have suffered from the disease, refers to the fact that this type of malaria, if uncomplicated, rarely results in the death of the patient.

   One of the most characteristic features of *P. vivax*, as seen in thin blood films, is the effect it produces on its host erythrocyte (Fig. 7.6a & b). As the parasite grows, the red cell becomes enlarged (to a much greater extent than could be due simply to mechanical pressure from the parasite) and pale in colour (probably due to the ingestion and

Table 7.1 Characteristics of species of *Plasmodium* which infect man.

Species	Geographical distribution	Duration of merogony — exoerythrocytic (days)	Duration of merogony — erythrocytic (hours)	Number of merozoites per meront — exoerythrocytic	Number of merozoites per meront — erythrocytic	Main morphological characteristics of erythrocytic forms — "ring form"	trophozoite	meront (diameter)	gametocytes	Host erythrocyte*
P. (P.) *vivax*	worldwide, in tropical, subtropical and warmer temperate regions	8	48	~ 10 000	12–24	at least one-third diameter of erythrocyte	ameboid	10 μm	round or ovoid; male, 9 μm; female, 10–11 μm	enlarged; stippled ("Schüffner's dots")
P. (P.) *malariae*	worldwide, but scattered, mainly tropical and subtropical	14–15	72	~ 15 000	6–12	at least one-third diameter of erythrocyte	compact, often bandlike	7 μm	round or ovoid, 7 μm	not enlarged; very faint ("Ziemann's") stippling after prolonged staining only
P. (P.) *ovale*	tropical Africa; also occasionally in other parts of tropics and subtropics (possibly of exogenous origin)	9	48	~ 15 000	6–12 (12–24 in relapses)	at least one-third diameter of erythrocyte	compact	7 μm (larger in relapses)	round or ovoid, 9 μm	slightly enlarged, stippled ("Schüffner's dots"); may be distorted and elongated
P. (L.) *falciparum*	worldwide, in tropics, subtropics, and warmer temperate regions	5½	48	~ 30 000	8–24	very small at first; 2 nuclei commonly; some apparently on edge of cell (accolé or appliqué forms)	compact; rarely seen in peripheral blood	5 μm (rarely seen in peripheral blood)	crescentic; male, 9–11 μm long; female, 12–14 μm	not enlarged; often with few large dots ("Maurer's clefts")

* These alterations do not develop until some growth of the parasite has occurred.

digestion of its hemoglobin by the parasite), and its surface membrane becomes covered with closely packed, fine dots, which stain pink with Giemsa's stain. The dots, named after their discoverer as Schüffner's dots, are not seen in unstained parasitized cells. Electron microscopy has shown Schüffner's dots to be small, complex vesicles or pits in the red-cell membrane, containing malarial proteins.

The parasite owes its specific name ("the lively *Plasmodium*") to the active ameboid movement shown by the growing trophozoite; thus in dried, stained blood films the trophozoites may be very irregular in shape.

Plasmodium (Plasmodium) malariae (Fig. 7.6h–l)   This species is less common than *P. vivax*. It holds the current record for longevity of infection in man, 40 years or more in untreated persons. Like *P. vivax*, it is not a very pathogenic organism, although chronic infections sometimes give rise to a lethal kidney condition. *P. malariae* is the only species of those commonly infecting man which also infects other primates. It is found in the West African chimpanzee. Since man and chimpanzees seldom live in close proximity, the importance of the latter as a reservoir of human disease is probably negligible. [The other species which infect man can develop exoerythrocytically in the livers of chimpanzees and other primates, but the blood is more or less resistant to infection unless the animal has had its spleen removed. Some strains of the *Plasmodium* species of man have been adapted to South American monkeys (*Aotus* and *Saimiri*) which are now used for experimental infections.]

Table 7.1 shows that both intra- and exoerythrocytic merogony are slower in *P. malariae* than in the other species, and this is true also of sporogony: at 24 °C, the latter cycle takes about 21 days in mosquitoes, compared with about 16 days for *P. ovale*, 11 days for *P. falciparum* and 9 days for *P. vivax*. *Plasmodium malariae* does not have as great an effect upon its host erythrocyte as does *P. vivax*. The red cell is not enlarged, and normally no stippling is seen. However, prolonged staining reveals very fine dots scattered somewhat irregularly over the plasmalemma (Ziemann's dots).

Plasmodium (Plasmodium) ovale (Fig. 7.6m–o)   This, the rarest of the four species which infect man, was not finally recognized as a distinct species until 1922, and not all workers accepted the distinction even then. Morphologically, it shows some resemblance to *P. malariae* (though the duration of erythrocytic merogony is different), while its effect on its host cell resembles that of *P. vivax*. It has been well described as "*P. malariae* in a *P. vivax*-type cell". The host cells show pronounced Schüffner's dots and are enlarged, though slightly less so than are erythrocytes infected with *P. vivax*. The infected erythrocytes seem to become unusually pliable, so that, in the making of a thin film, they may be drawn out into an elongated oval shape (hence the specific name) and even show a tattered and torn appearance at one end. *Plasmodium ovale*, like the two preceding species, is not greatly pathogenic.

Plasmodium (Laverania) falciparum (Fig. 7.6p–x)   This parasite has the doubtful distinction, according to P. C. C. Garnham, of being "almost unchallenged in its supremacy as the greatest killer of the human race over most parts of Africa and elsewhere in the tropics". The pathology of the disease (malignant tertian malaria) caused by it is described below. The specific name is derived from the fact that, in this species, the gametocytes are crescentic (Fig. 7.6v, w, x).

The growing trophozoites, meronts and immature gametocytes of *P. falciparum* are very rarely seen in the circulating blood, as they are concentrated within the capillaries and blood sinuses of internal organs such as the brain, liver, kidneys, spleen, bone-marrow, and especially – in pregnant women – the placenta. The reason for this seems to be that the membranes of infected erythrocytes develop "knobs" which may cause them to adhere to the walls of the smaller vessels. Thus blood films made from persons infected with *P. falciparum* usually contain only very young trophozoites ("ring forms") and mature gametocytes.

The erythrocytes containing asexual parasites, except for those inhabited by the very young ring forms, show, when stained, a few, relatively large dots or lines called Maurer's clefts, but no Schüffner's or Ziemann's dots. Electron microscopy has shown that these clefts are within the cytoplasm of the erythrocyte and are extensions of the original parasitophorous vacuole.

Pathology of malaria in man

It has already been stated that only *P. falciparum* directly causes fatal disease in man. The other three species, if untreated, cause recurrent fevers which are very debilitating and may lower the patient's resistance to other infection, but they do not usually, of themselves, kill.

The fevers, so characteristic of malaria, occur when the meronts in the erythrocytes burst, setting free their merozoites and also liberating into the blood various secretory and excretory products, including antigenic capsular material and the malarial pigment as well as the remains of the red cell cytoplasm. Some of these liberated substances are immunogenic and may induce an allergic reaction which is the cause of high fever, and the subsequent rigors (violent shivering). The development of the meronts tends to be synchronous, so that they will all burst at the same time. The time elapsing between successive fevers is therefore often the same as the duration of the erythrocytic merogonic cycle, i.e. 72 hours with *P. malariae* and 48 hours with the other three species. Hence, the fevers produced by *P. malariae* are said to be "quartan", unlike those of the other three species, which are "tertian". Frequently, during the first few days of an attack, merogony is less synchronized, and fevers may occur daily; but usually they soon adopt the rhythm characteristic of the species.

Exoerythrocytic meronts of mammalian malarial parasites (though not those of all avian species) are entirely nonpathogenic. No immune response develops against the exoerythrocytic stage of mammalian species of *Plasmodium* under natural conditions. Thus, during the early part of the infection in mammals, before the erythrocytes are invaded, no symptom develops. This is called the prepatent period (the period before parasites can be detected in the blood). The period between infection and the appearance of the first symptoms is the incubation period. It clearly cannot be shorter than the prepatent period, and must be longer by the duration of at least one erythrocytic cycle since symptoms do not develop until the first erythrocytic meronts burst. If the infection is slight, the incubation period may be still longer, since the presence of only a few rupturing meronts in the blood may not produce noticeable symptoms.

If the infection is not treated and if death does not result rapidly (as it may with *P. falciparum*), after a variable number of erythrocytic merogonic cycles the host's immune response will bring the erythrocytic infection under control. All symptoms will disappear and eventually parasites will no longer be detectable in blood samples: the infection is now latent. However, cure will not be complete: either a few erythrocytic meronts survive in the capillaries of various viscera or hypnozoites persist in the liver. Eventually, when either the level of immunity has dropped or when the parasite itself has changed its antigenic structure (perhaps in the way some trypanosomes do) so that it is no longer affected by the existing immune response, the number of parasites in erythrocytes increases and a second clinical attack develops. When this recurrence results from erythrocytic forms, it is (strictly) termed a "recrudescence"; when from hypnozoites, it is a true "relapse". In the absence of treatment, remission and relapse (or recrudescence) may continue for many years. Persons living in areas where malaria is common are regularly reinfected throughout their lives by the bite of infected mosquitoes and may, if they survive, develop complete resistance to, or partial tolerance of, the parasites; however, the infant death rate in such areas is appallingly high unless

175

medical facilities are adequate to permit rapid diagnosis and treatment.

Apart from the fever, malarial infection (in man and other animals) always leads to a massive increase in the number of phagocytic cells. The spleen is the largest agglomeration of these cells, and so it becomes grossly enlarged in individuals with chronic malaria. The spleen seems to be of great importance in the body's defence against malaria. In experimentally infected animals, in which the infection has become latent, surgical removal of the spleen leads to the rapid reappearance of parasites in the blood (compare its effect in human infections with *Babesia divergens*, Ch. 8).

Plasmodium falciparum exerts its lethal effect mainly by causing

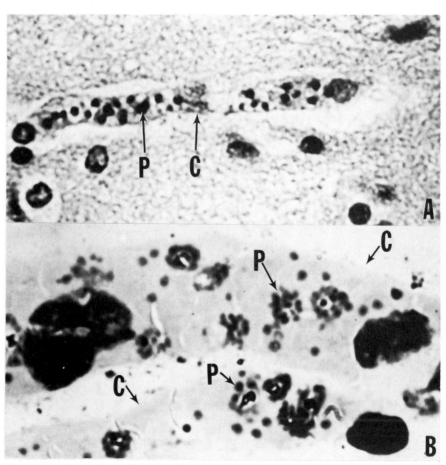

Figure 7.8 *Plasmodium falciparum* (P) in blood cells within capillaries (C) of the brain. (a) At this relatively low magnification the individual meront nuclei in the parasites cannot be distinguished. (b) At this high magnification the meront nuclei are individually visible, as are the host erythrocytes.

blockage of the capillaries in the brain (Fig. 7.8) and other organs. The precise mechanism of this effect is not fully understood, but it seems to be due partly to the fact that the surface membranes of infected erythrocytes develop the "knobs" referred to above, which act as receptors for the endothelial lining of the blood vessels so that the infected erythrocytes adhere to the capillary walls, and partly to the development of an inflammatory response to the parasite and host cell components released at merogony. This leads first to interference with the tissue's oxygen supply, and eventually to rupture of the blocked capillaries and bleeding into the surrounding tissue. This may occur in any or all of the internal organs, but is most serious when it occurs in the brain, and it is damage to this organ that causes the death of most persons who die from acute malignant tertian malaria ("cerebral malaria").

Another very serious complication of *P. falciparum* is blackwater fever. This results from large-scale lysis of erythrocytes of patients with *P. falciparum* malaria and the consequent excretion of altered hemoglobin in the urine, giving the latter its dark color for which the condition was named. Blackwater fever has long been known to be associated with quinine treatment. It is now thought that the condition may result from the fact that free fatty acids in the plasma are hemolytic, but that this effect is countered by plasma proteins which bind to the fatty acids and render them harmless. Fatty acids are particularly abundant in bloods of people infected with *P. falciparum*. Quinine may interfere with the protective effect of the protein, and wholesale lysis of red cells may result.

Diagnosis of malaria in man

Apart from the clinical findings (recurrent fevers, enlarged spleen), diagnosis depends on the demonstration of parasites in thick (Fig. 7.9) or thin blood films (Fig. 7.6). There are no suitable experimental animals into which blood can be injected to detect scanty infections. Serological procedures such as fluorescent antibody tests and enzyme-linked immunosorbent assay (ELISA) are increasingly used in doubtful cases.

Treatment and prevention of malaria in man

The drug quinine, an extract of the bark of the cinchona tree, was known for centuries in Peru before being imported into Europe during the first half of the seventeenth century. It is very effective in rapidly destroying the erythrocytic parasites but, because of its association with blackwater fever, treatment is now more usually given by

177

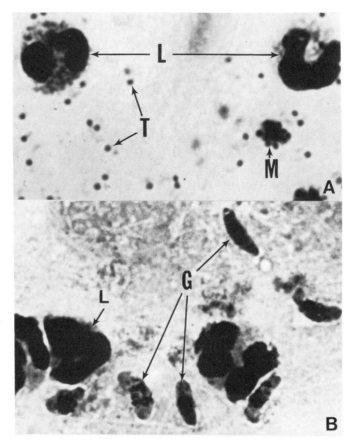

Figure 7.9 Thick films of blood infected with *P. falciparum*. In (a) trophozoites (T) and a meront (M) are present. In (b) the crescent-shaped gametocytes (G) are visible. Leukocytes (L) are present in both blood films.

synthetic drugs. Within the last few decades, however, an increasing resistance to synthetic drugs, especially chloroquine, on the part of malaria parasites (*P. falciparum* in particular) has led to a return to quinine as a last resort. Some degree of drug resistance has been reported from many regions of Africa and South America, and much of the Far East including parts of China, India, and Bangladesh. The problem of drug resistance is spreading, and this causes great concern.

Rapid treatment of an acute clinical attack of malaria is best achieved by drugs which attack the erythrocytic meronts. Such drugs include quinine and chloroquine, a member of a group of compounds called 4-aminoquinolines. Unfortunately, resistance to chloroquine usually also confers some degree of resistance to other 4-aminoquinolines, such as amodiaquine, which could otherwise be used in treatment. There are also indications that amodiaquine may produce unwanted side effects if given together with proguanil. In areas where resistance does not occur, chloroquine remains the best drug for treatment. Elsewhere, a combination of pyrimethamine (2:4-diaminopyrimidine) with sulphadiazine (a sulphonamide), known as "Fansidar", is used, but recently resistance even to that has been reported. Where this occurs, a newly developed compound, chemically related to quinine, called mefloquine ("Lariam") can be used or a triple combination of mefloquine, pyrimethamine and sulphadiazine ("Fansimef"). However, neither "Fansidar" nor "Fansimef" is suitable for use by patients allergic to sulphonamides, and neither it nor mefloquine is advised for treatment of pregnant women or infants. A herbal extract developed in China, called *qinghaosu* (or artemisine), and its derivatives, are under trial and looks promising. *Qinghaosu* is unrelated chemically to any of the other antimalarial drugs and therefore resistance to it in strains of *Plasmodium* resistant to other drugs is not likely.

Prevention of clinical malaria can be achieved by the use of drugs which attack either the exoerythrocytic stage (causal prophylactics) or the erythrocytic parasites, thus preventing clinical symptoms from appearing (chemosuppressants). Causal prophylactics include proguanil ("Paludrine") and the related chlorproguanil – the former being taken daily and the latter, weekly. Resistance to proguanil is becoming a problem, though it is less widespread than is resistance to chloroquine.

As suppressants, chloroquine or pyrimethamine given once a week were widely used. Use of chloroquine is now discouraged due to the danger of exacerbating the development and spread of resistance to the drug. There are indications of the development of resistance to pyrimethamine in some areas, so that this drug also is not recommended. Weekly amodiaquine is perhaps the best compromise suppressant regimen, and its use for this purpose was recommended by the World Health Organization in 1985.

"Fansidar" (see above) can be used if proguanil resistance is a problem. Mefloquine, another double combination, "Maloprim" (dapsone and pyrimethamine), or the triple combination "Fansimef", can also be used as short-term suppressants. The safety of these drugs has not yet been evaluated adequately and thus long-term use is not recommended.

The use of suppressants (as distinct from causal prophylactics) does not prevent development of hypnozoites in *P. vivax* infections. Thus relapses may occur when the suppressant regimen is discontinued – as they may also following clinical "cure" of *P. vivax* infection with chloroquine, quinine, mefloquine or "Fansimef." To prevent this, treatment with the tissue meronticide, primaquine, is necessary. Primaquine, an 8-aminoquinoline, is too toxic for routine use as a prophylactic and it is best only to administer it to patients in hospital.

The development of an antimalarial vaccine now seems a possibility within the foreseeable future. Antisporozoite vaccines seem, at the time of writing, to be the most likely candidates for success. Their development is advanced and as sporozoites appear to be unable to undergo antigenic variation, their use would avoid the problems of antigenic variation which arise with vaccines against the intraerythrocytic stages.

Apart from immunoprophylaxis or chemosuppression, malaria can, of course, be prevented by avoiding contact with infective mosquitoes. Since all the vector species of *Anopheles* feed in the evening or at night, this can be done by screening the windows and doors of houses with fine mesh netting and by the use of insecticidal sprays and mosquito nets over beds. Also, if adequate funds are available, the mosquitoes can be destroyed, and malaria eradicated, by spraying houses with long-lasting insecticides and by the drainage of swamps, and other wet areas where they breed. Thanks to the efforts of local health authorities, governments and the World Health Organization, much progress has been made in eliminating malaria from towns, districts, and even whole countries where it was previously a scourge; but much still remains to be done and mosquitoes are, regrettably, becoming increasingly resistant to many insecticides.

Other species of Plasmodium

Many species of *Plasmodium* which infect animals other than man are used for experimental work. Some of the more important of these are listed below.

Plasmodium (Plasmodium) cynomolgi   This species infects monkeys (mainly *Macaca* spp.) in India, Ceylon, and the Far East. It is very similar to *P. vivax*, and several subspecies exist, at least some of which can also infect man.

Plasmodium (Plasmodium) knowlesi   Another parasite of *Macaca* spp. in Asia, causing a fatal disease, this species is unique among those infecting primates in having a 24-hour erythrocytic schizogony cycle. It is lethal to rhesus monkeys and can infect man accidentally or experimentally, but causes only mild disease.

Several other species have been described from Asian and South American monkeys. In Africa only one species is known in monkeys,

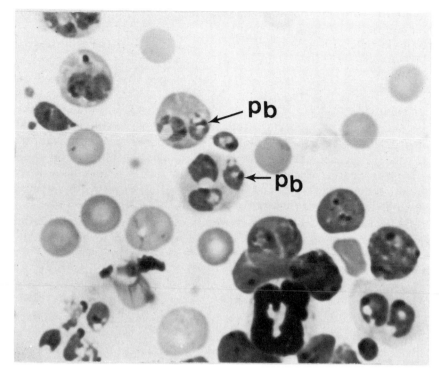

Figure 7.10 *Plasmodium berghei* (Pb) in blood film from a rat. *P. berghei* infections are not synchronized so organisms in all stages of development may be present.

P. (P.) gonderi, but others infect the higher apes (chimpanzees and gorillas). Usually they are not markedly pathogenic.

Plasmodium (Vinckeia) berghei (Fig. 7.10)   One of a complex of species of malaria parasitic in African murine rodents, *P. berghei* was discovered in Zaire (then the Belgian Congo). It can be maintained in laboratory rats and mice, and is fairly pathogenic to them. Other members of this group of murine parasites, increasingly used in laboratory studies, are *P. yoelii*, *P. vinckei* and *P. chabaudi*.

Plasmodium (Haemamoeba) gallinaceum (Fig. 7.5)   A natural parasite of the jungle-fowl of Asia, this species is commonly maintained in laboratories and is sometimes used in testing possible new antimalarial drugs. It can cause outbreaks of disease in flocks of domestic hens and often kills the younger birds as a result of blockage of the brain capillaries by the large secondary exoerythrocytic schizonts, which develop in the endothelial cells (Fig. 7.5c).

Most of the species of *Plasmodium* which infect birds and reptiles do not seem to be very pathogenic.

FAMILY HAEMOPROTEIDAE

Members of this family are very similar to *Plasmodium* in their morphology and life cycles, but differ in not having erythrocytic schizogony and in being transmitted by insects other than mosquitoes. The gametocytes, which are the only stage to be found in erythrocytes, contain pigment and cannot be distinguished at the generic level from the gametocytes of *Plasmodium*. Exoerythrocytic meronts of members of this family are found in various tissues, and are sometimes quite large. Haemoproteids have been described from mammals, birds, and reptiles, and are grouped into seven genera (listed below). Most species are only slightly, if at all, harmful to their hosts.

Hepatocystis *Levaditi and Schoen, 1932*

This is the commonest hemoproteid of mammals. The exoerythrocytic meronts, in liver parenchyma cells, are very large (up to 1 mm) and are called merocysts. The life cycle of only one species (*H. kochi* of African monkeys) is known; it is transmitted by *Culicoides* spp. (midges: Diptera, family Ceratopogonidae).

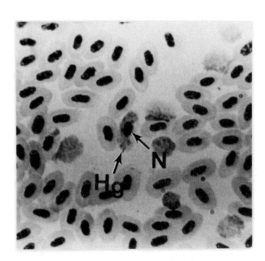

Figure 7.11 *Haemoproteus* sp. in bird blood. The hemoproteid gametocyte (Hg) almost surrounds the host cell nucleus (N).

Polychromophilus *Dionisi, 1899 and*
Nycteria *Garnham and Heisch, 1953*

These organisms infect insectivorous bats (Chiroptera) in many countries.

Haemoproteus *Kruse, 1980*

A common parasite of birds throughout the world. Several species have been described. *Haemoproteus palumbis* (Fig. 7.11) occurs frequently in English wood pigeons (*Columba palumbus*), and *H. columbae* occurs in the related *C. livia* in the USA and elsewhere. *Haemoproteus* is transmitted by large ectoparasitic flies of the family Hippoboscidae ("louse-flies": Diptera), and merogony occurs chiefly in the lung.

Parahaemoproteus *Bennett, Garnham and Fallis, 1965*

Another common genus found in birds, including ducks in North America. Merogony is mainly in viscera other than the lung, and the vectors, where known, are midges (*Culicoides*).

Haemocystidium *Castellani and Willey, 1904 and*
Simondia *Garnham, 1966*

These organisms infect only reptiles.

FAMILY LEUCOCYTOZOIDAE

Members of this family are known only from birds, with one exception described from a reptile. As in the Haemoproteidae, only gametocytes are found in blood cells. Originally they were thought to inhabit leukocytes (hence the generic name), but now they are known, in at least some species, to infect the precursors of the erythrocytes. Infected cells become enlarged, and their nuclei are considerably altered, so that only while the contained parasite is very young is it possible to identify the host cell. Infection with certain species results in a curious but characteristic elongation of some of the host cells, which become spindle-shaped; the contained gametocyte is elongated (Fig. 7.12a). Other gametocytes (probably produced by a different type of meront) and their host cells are rounded (Fig. 7.12b). The gametocytes of all species in this family are larger than those of the Haemoproteidae and Plasmodiidae, and differ in not having any malarial pigment, probably because they invade cells in which hemoglobin has not yet been

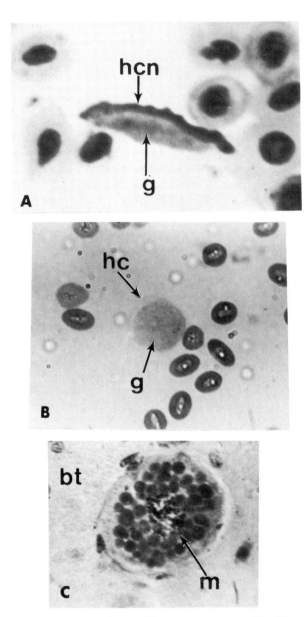

Figure 7.12 *Leucocytozoon simondi.* (a) Elongate gametocyte (g). The elongated dark structure is the remnant of the host cell nucleus (hcn). (b) Round gametocyte (g) within a host cell (hc); no host cell nucleus is present in this cell. (c) Meront (m) in brain tissue (bt).

formed. The life cycle of leucocytozoids is basically similar to that of the other two families in the Haemosporina. Exoerythrocytic meronts are usually large (Fig. 7.12c), and may occur in various internal organs. Three genera are recognized.

Leucocytozoon *Ziemann, 1898*

Many species have been described from birds the world over. Most are seemingly harmless, but one, *L. simondi* (Fig. 7.12), causes severe disease and death in domestic ducks in North America, while wild ducks serve as a reservoir of infection and, as is so often true, seem to be less severely infected. Wild birds in many countries (including the USA and Britain) are infected with various species of *Leucocytozoon*. Where known, the vectors are blackflies of the genus *Simulium* (Diptera, Nematocera, Simuliidae).

Akiba *Bennett, Garnham and Fallis, 1965*

This genus contains a single species, *A. caulleryi*, which causes disease in chickens in Japan and eastern Asia. It was previously regarded as a species of *Leucocytozoon*, but was reclassified largely on the grounds of its transmission by *Culicoides* instead of *Simulium*; whether this separation was justified remains to be determined.

Saurocytozoon *Lainson and Shaw, 1969*

This genus, with a single species, was described from a Brazilian lizard in 1969. Morphologically it resembles a species of *Leucocytozoon* with round gametocytes.

FURTHER READING

Bannister, L. H. & R. E. Sinden (1982). New knowledge of parasite morphology. *British Medical Bulletin* 38, 141–5.
Bruce-Chwatt, L. J. 1984. *Essential malariology*, 2nd edn. London: Heinemann Medical Books.
Coatney, C. R., W. E. Collins, McW. Warren & P. Contacos 1971. *The primate malarias*. Bethesda, Maryland: National Institute of Health.
Cohen, S. (ed.) 1982. Malaria. *British Medical Bulletin no. 2*, 103–99.
Garnham, P. C. C. 1966. *Malaria parasites and other Haemosporidia*. Oxford: Blackwell Scientific.
Garnham, P. C. C. 1984. Life cycles. In *Handbook of experimental pharmacology*, Vol. 68/I, W. Peters & W. H. G. Richards (eds.), 3–30. Berlin: Springer-Verlag.

Killick-Kendrick, R. & W. Peters (eds) 1978. *Rodent malaria*. London: Academic Press.

Kolata, G. 1984. The search for a malaria vaccine. *Science* **226**, 679–82.

Kreier, J. P. (ed.) 1980. *Malaria*, 3 vols. New York: Academic Press.

Laser, H., P. Kemp, N. Miller, D. Lander & R. Klein 1975. Malaria, quinine and red cell lysis. *Parasitology* **71**, 167–81.

Maddox, J. 1984. Malaria vaccine in sight? *Nature* **310**, 541.

Peters, W. 1970. *Chemotherapy and drug resistance in malaria*. London: Academic Press.

Phillips, R. S. 1983. *Malaria*, Studies in biology 152. London: Edward Arnold.

Ristic, M., Ambroise-Thomas & J. P. Kreier 1984. *Malaria and babesiosis*. Dordrecht, The Netherlands; Martinus Nijhoff.

Sinden, R. E. 1983. Sexual development of malarial parasites. *Advances in Parasitology* **22**, 153–216.

Walliker, D. 1983. The genetic basis of diversity in malaria parasites. *Advances in Parasitology* **22**, 217–59.

CHAPTER EIGHT

Piroplasms

These organisms can be simply defined as symbiotic protozoa of which the trophic forms inhabit erythrocytes, and sometimes other cells, of vertebrates but do not form pigment from hemoglobin. As far as is known, they are all transmitted by ticks (Arthropoda, Acarina; families Ixodidae and, for one species of *Babesia* only, Argasidae). However, other vectors may remain to be discovered, particularly for those piroplasm whose hosts are aquatic. All the piroplasms are small, usually round or pear-shaped when in their host's erythrocytes (hence the name piroplasm). They are known from fish, amphibia, birds, and mammals. None of them naturally infects man, though occasional accidental infections have been recorded.

Piroplasms are now classified (after years of debate) in the phylum Apicomplexa, as a subclass, Piroplasmia. Within the subclass, all known genera are grouped in a single order Piroplasmida, which is divided (by many but not all authorities) into three families: Babesiidae, Theileriidae, and Dactylosomidae.

FAMILY BABESIIDAE

Most members of this family parasitize mammals; a few infect birds or reptiles. Intraerythrocytic stages reproduce by binary fission or budding into two (rarely four) individuals. Until relatively recently (1980) it was thought that babesiids infected only the erythrocytes of their vertebrate hosts. It is now known that at least two species (*B. microti, B. equi*) have a single pre-erythrocytic generation of meronts within lymphocytes; perhaps these species should be transferred to another genus. Other generic names which are sometimes, or have been, given to some babesiids are *Nuttallia, Nicollia,* and *Piroplasma* (the last being belatedly recognized as an invalid junior synonym of *Babesia*). For convenience, we shall refer all babesiids to one genus, *Babesia*.

187

Table 8.1 Babesiidae of veterinary importance.

Species	Size*	Geographical distribution	Main hosts		Patho-genicity	Common name of disease (if any)
			vertebrate	invertebrate		
Babesia bigemina	large	Central & South America, Europe, Africa, Australia	cattle, deer	*Boophilus, Haemaphysalis, Rhipicephalus*	moderate	redwater fever
B. bovis (*B. argentina*)	small	Europe, Russia, Africa, Australia, South & Central America	cattle, deer	*Ixodes, Boophilus, Rhipicephalus*	high	redwater fever
B. divergens	small	Western & Central Europe including England	cattle	*Ixodes*	moderate	redwater fever
B. major	large	Europe, Russia	cattle	*Boophilus*	low	
B. caballi	large	Southern Europe, Asia, Russia, Africa	Equidae (domestic)	*Dermacentor, Hyalomma, Rhipicephalus*	moderate	
B. equi	small	Southern Europe, Asia, Russia, Africa, South America	Equidae (domestic and zebra)	*Dermacentor, Hyalomma, Rhipicephalus*	high	biliary fever

B. motasi	large	Southern Europe, Russia, Africa, Far East, tropical America	sheep, goats	*Rhipicephalus, Haemaphysalis, Dermacentor*	moderate	
B. ovis	small	As for *B. motasi*	sheep, goats	*Rhipicephalus, Ixodes*	low	
B. trautmanni	large	Southern Europe, Africa, Russia	pig	*Rhipicephalus*	moderate	
B. canis	large	North, Central, South America, Southern Europe, Russia, Africa, Asia	dogs and wild canids	*Rhipicephalus, Dermacentor, Haemaphysalis*	moderate–high	tick fever
B. gibsoni	small	Indian, Ceylon, China	dogs and wild canids	*Rhipicephalus, Haemaphysalis*	high (in domestic dog)	tick fever
B. felis	small	Africa, India	domestic cat, lion, leopard	*Haemaphysalis?*	moderate	
B. microti†	small	World-wide	rodents	*Ixodes*	low	
B. rodhaini†	small	Africa	rodents	unknown	low	

* In this Table, "large" means about 2–5 μm long, and "small" means about 1–2 μm long.
† These species, while not strictly speaking of veterinary importance, have been domesticated and are maintained in hamsters (*B. microti*) and rats and mice (*B. rodhaini*) in laboratories around the world. They are widely used for preliminary screening of drugs for antibabesial activity and have thus assumed veterinary importance.

GENUS *BABESIA* STARCOVICI, 1893

Some of the commoner species of *Babesia* are of veterinary importance, their main hosts and geograpical distribution are listed in Table 8.1. A large number of species of *Babesia* has been described from wild rodents and other wild animals. Many of these appear to be quite benign in their hosts. Some *Babesia* species produce disease (babesiosis) in domestic animals. Babesiosis of cattle, horses, and dogs is often particularly severe. *Babesia rodhaini*, a natural parasite of wild rodents in Africa, and *B. microti*, a world-wide parasite of rodents, are often maintained in laboratory mice for use in experimental work. *Babesia microti* has been a cause of human infection, mainly in the New England and New York coastal regions of the United States; it is one of the species (noted above) with a pre-erythrocytic cycle of merogony within lymphocytes.

Morphology and life cycle

Babesia develop in the vertebrate host in erythrocytes (and, as noted earlier, in some species in lymphocytes also). In the arthropod vector *Babesia* develop in a variety of cells and tissues. *Babesia* species are classed as large (2–5 µm long) and small (1–2 µm long) types. Many hosts may be infected with both a large species and a small one. The large and small species differ to some extent in susceptibility to drugs (e.g. large species are susceptible to trypan blue; small are not) and in mode of reproduction in erythrocytes (e.g. large species produce one or two merozoites at a time, and small ones may produce four – yielding a "Maltese cross" form). To date, development in lymphocytes has been reported for small species only. The generally neglected proposal that the small species, dividing into four in erythrocytes, be grouped into the genus *Nuttallia* is perhaps strengthened by the demonstration of extra-erythrocytic development in some members of the group.

In erythrocytes both large and small species exist as round or oval organisms (Fig. 8.1). Division stages are often seen (Fig. 8.2). The products of division are merozoites possessing a typical apical complex (but no conoid).

The life cycle in the tick vector is very complicated and has only recently been elucidated. The great delay in elucidation of the tick phases of the organism is surprising in light of the fact that a species of *Babesia* was the first parasite shown to be transmitted by an arthropod, by Smith & Kilborne in 1893.

Some species of tick feed only once during each stage (larva, nymph,

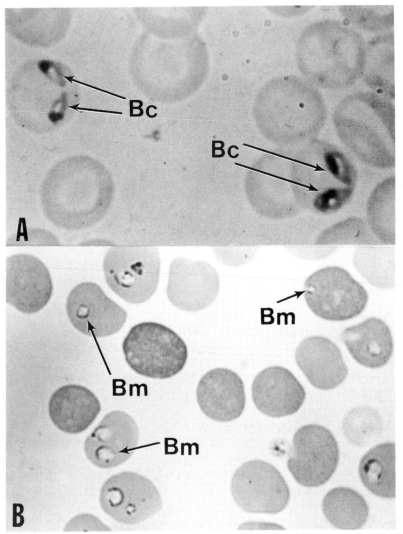

Figure 8.1 Photomicrograph of *Babesia canis* (a) and *B. microti* (b) in thin blood films. The *Babesia canis* (Bc) organisms appear in the paired form typical of the large species of *Babesia*. *B. microti* (Bm) is among the larger of the small species and frequently appears as rings somewhat resembling the ring-form trophozoites of *Plasmodium*.

adult) of their life cycle (Fig. 8.3), and in these species the piroplasms are passed from stage to stage so that if such a tick, for example, becomes infected as a larva, it will transmit the infection only after molting and becoming a nymph. Some babesiids may also enter the eggs of adult female ticks and so pass to the larvae before being transmitted to another vertebrate.

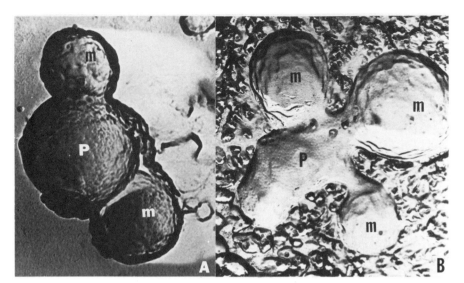

Figure 8.2 Electron micrographs of *Babesia microti* in the process of reproduction. In (a) the parent organism (P) is budding one or possibly two merozoites (m). In (b) the parent organism (P) is budding three merozoites (m). The organism in (b) would appear as a "Maltese cross" in a stained blood film. (From Kreier *et al.* 1975.)

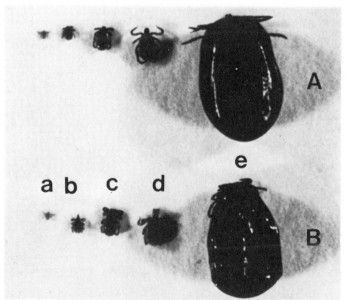

Figure 8.3 *Ixodes daminii* (a) and *I. pacificus* (b). These ticks are vectors of *Babesia microti*. Transmission of this parasite is trans-stadial. Infection is acquired by the larva (a) and the protozoan is passed to the next host by the nymph (b). The adult male (c) does not engorge but the female (d) ingests so much blood that she becomes a blood-filled sac (e).

The occurrence of sexual reproduction in the life cycle of babesiids, long the subject of debate, has now been confirmed. Some of the intraerythrocytic forms are gamonts. In the vector tick, the gamonts leave their host cells and transform into spikey "ray-bodies" or "Strahlenkörper". Strahlenkörper were described by Robert Koch many years ago but their existence was later disputed. Two ray-bodies, presumably male and female gametes, fuse to form a zygote which becomes a motile kinete (equivalent to the malarial ookinete). The kinete leaves the intestine and penetrates cells of various tissues – hemocytes, muscles, and Malpighian tubules. In those species of

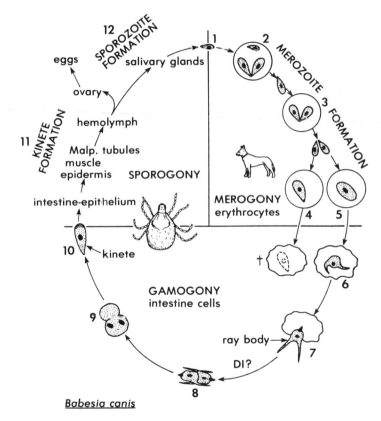

Figure 8.4 Diagrammatic representation of the life cycle of *Babesia canis*. (1) Sporozoite in saliva of feeding tick. (2–3) Asexual reproduction in erythrocytes of vertebrate host (dog) producing merozoites, (4) which are lysed(†) in the gut of the tick. (5–6) In the tick's intestine certain ovoid intraerythrocytic stages are not lysed but form protrusions and thus become gametes ("ray-bodies"). (7–8) Two ray-bodies fuse with each other; DI? indicates possible division(s) at this stage. (9–10) Formation of a single motile kinete from a zygote. (11) After leaving the intestine the kinetes enter various organs and initiate formation of new kinetes. In *B. canis*, eggs are penetrated and the next generation of ticks is infected. (12) A kinete that has penetrated salivary glands gives rise to thousands of small sporozoites (from H. Mehlhorn & E. Schein 1984).

Babesia in which transovarial transmission occurs, some of the kinetes enter eggs and infect the next generation of ticks, thus accounting for adult to larva transmission. The intracellular kinetes reproduce by multiple fission to form more kinetes (sporokinetes), which reinvade other cells and may repeat the process or, if they enter salivary gland cells, become sporonts and divide by either multiple or repeated binary fissions to form sporozoites. Sporozoite development begins only after the infected tick has attached to a new host, and is then completed rapidly (within five days) so that the infective sporozoites can be injected into the new host before the tick finishes feeding on it. The sporozoites enter erythrocytes (or lymphocytes, in the case of *B. microti* or *B. equi*) and recommence the life cycle. The life cycle of *B. canis*, which may be considered representative, is shown in Figure 8.4.

Pathogenesis

When species of *Babesia* are pathogenic, the disease is usually associated with anemia, fever, and enlargement of the spleen (which is not dark in color, as it is in malaria, because of the absence of malarial pigment in babesiosis). Blocking of the capillaries in various tissues (including the brain), which may damage the cells by depleting their oxygen supply (as in malaria due to *P. falciparum*), occurs during *B. bovis* and *B. equi* infection. In *B. bigemina* infection of cattle in particular, the anemia is often accompanied by lysis of erythrocytes and excretion of the released hemoglobin in the urine; hence, babesiosis of cattle is known as redwater fever. Animals which recover from an acute attack usually become tolerant, not becoming clinically ill so long as they remain chronically infected, which may be for two years or more if they are constantly being bitten by infected ticks. Relapses are not known to occur.

Diagnosis

Infections are diagnosed from the clinical signs and symptoms, if any, and by finding organisms in blood films (Fig. 8.1). In animals with acute infections parasites may usually be identified in stained thin blood films. In some animals piroplasms may be sparse, and repeated and careful examination may be required to detect them. A variety of serological tests is used to aid in detection of carrier animals which may have very low parasitemias. These include complement fixation tests, fluorescent antibody tests and ELISA for detection of antibody to the organisms. A test for soluble babesial antigen in the serum has been proposed for detection of early infections.

194

Treatment and prevention

A variety of drugs is used in treating babesiosis of domestic animals, including some of those active against trypanosomes (e.g. "Berenil", diminazene). Others used include trypan blue (particularly effective against large babesiids), acaprin, amicarbalide, and acriflavine. Commonly used antimalarials, such as chloroquine, are not effective against *Babesia*.

Babesiosis of domestic animals can best be prevented by the use of insecticides and good husbandry to keep animals free from ticks.

Human babesiosis

Although not normally infective to man, human infections with babesiids have been recorded. The first of these was in 1956, in a Jugoslav farmer whose spleen had been removed after an accident 11 years earlier: he developed a fatal infection with *Babesia divergens*. The second case was an American hunter, who in 1966 became infected with an unidentified babesiid. Subsequent infections in America have been shown to be due to *B. microti* and are usually self-limiting. Those in Europe were all caused by *B. divergens*; they occurred exclusively (unlike the American cases) in splenectomized persons and most ended fatally.

Babesia may be difficult to distinguish from "ring" forms of *Plasmodium* species in thick films and even in thin films. However, if division forms are seen this confusion should not occur, as malarial meronts are almost always considerably larger, and produce more daughter individuals, than dividing piroplasms; they also contain pigment.

About 100 human infections with *Babesia* are known, though additional infections may have occurred and have been misdiagnosed as malaria, or missed because they were asymptomatic. Asymptomatic cases have been identified in Mexico. Thus, the infection may be commoner than is at present generally supposed.

There is no known safe and effective chemotherapy for human babesiosis, though quinine plus the antibiotic clindamycin, and diminazene, have been tried, and pentamidine has been suggested. Physicians treating patients infected with *Babesia* could perhaps draw on veterinary literature for guidance.

FAMILY THEILERIIDAE

The forms inhabiting erythrocytes are very small (1–2 µm). Members of this family were historically separated from the babesiids by the fact that in the vertebrate they undergo merogony in other cells (lymphocytes) in addition to erythrocytes. This distinction is now questionable with the demonstration that at least some species of *Babesia* (i.e. *B. microti* and *B. equi*) have stages in lymphocytes.

The type and location of exoerythrocytic merogony is used as the basis for dividing the family into two genera: *Theileria*, with small or medium-sized (about 10–20 µm in diameter) exoerythrocytic meronts in lymphocytes, and *Cytauxzoon*, with large exoerythrocytic meronts in phagocytic cells in the vascular endothelium. The meronts are over 50 µm diameter when mature, and are divided into cytomeres.

The accuracy of this simple classification is in some doubt and it may need to be revised. A proposal to split the genus *Theileria* into two genera, *Theileria* and *Gonderia*, on the basis of the supposed absence of reproduction of the erythrocytic piroplasm in some species (i.e. *T. parva*) has not been generally accepted.

The theileriids infect ruminants (and, of course, their tick vectors), mainly but not exclusively in the tropics. *Theileria* often seriously affects domestic animals such as cattle, sheep, and pigs, which are imported into enzootic areas from areas free of the infection.

GENUS *THEILERIA* BETTENCOURT, FRANÇA AND BORGES, 1907

The important species of this genus are listed in Table 8.2, together with their main hosts and geographical distribution. In gross morphology, the erythrocytic forms (Fig. 8.5) resemble small babesiids, about 1–2 µm long by 0.5–1 µm wide. It is not always easy to distinguish them in a blood film from some of the smaller babesiids. *Theileria* in erythrocytes lack an apical complex, but this can, of course, be determined only by electron microscopy.

The exoerythrocytic meronts are found chiefly in the spleen and lymph nodes, though they may also be seen wherever there are lymphocytes (i.e. in all viscera and in the blood). They are irregular in shape, usually round or oval, and are about 10–20 µm in diameter when fully grown (Fig. 8.6). They contain a mass of cytoplasm with a number of nuclei, and finally bud off many small uninucleate merozoites which invade the erythrocytes. When an infected lymphocyte divides, both daughter cells appear to retain part of the meront and so

Table 8.2 Theileriidae of veterinary importance.

Species	Geographical distribution	Main hosts		Patho-genicity	Common name of disease (if any)
		vertebrate	invertebrate		
Theileria parva	East Africa (eradicated from southern Africa)	cattle, buffalo	*Rhipicephalus, Hyalomma*	high	East coast fever
T. annulata	North Africa, southern Europe, southern Russia, India, China	cattle, water buffalo	*Hyalomma*	high	Mediterranean coast fever
T. mutans	Africa	cattle	*Amblyomma*	low–moderate	
T. sergenti	East Asia, India, Eastern USSR, Japan	cattle	*Haemaphysalis*	moderate	
Theileria (species undefined; was called *T. mutans*)	world-wide	cattle	*Haemaphysalis*	low	
T. hirci	North Africa, south-east Europe, southern Russia, Middle East	sheep, goat	*Rhipicephalus?*	high	
T. ovis	Africa, Europe, Russia, India, Middle East	sheep, goat	*Rhipicephalus*	very low	
Cytauxzoon felis	North America	cat, lynx	Unknown	high in cat	

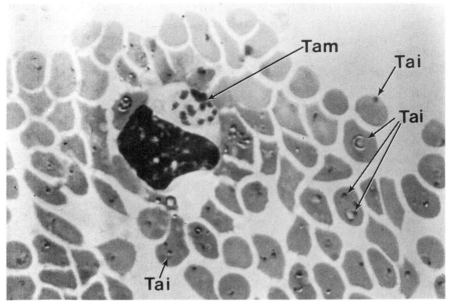

Figure 8.5 Photomicrograph of *Theileria annulata* in thin blood film from a cow. In this film intraerythrocytic forms (Tai) and a meront in a lymphocyte (Tam) are present (the blood film from which this micrograph was made was kindly provided by Dr O. P. Gautam, Haryana Agricultural University).

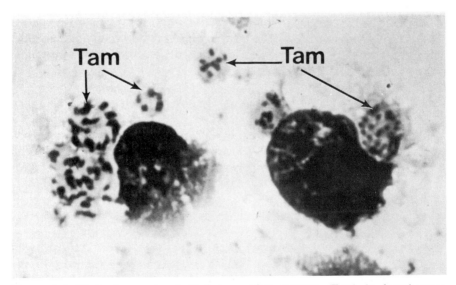

Figure 8.6 Photomicrograph of *Theileria annulata* meronts (Tam) in lymphocytes obtained from a lymph node. Some meronts are within the cytoplasm of the lymphocytes and others have been released as a result of disruption of lymphocytes during sample collection and preparation.

the number of infected lymphocytes increases. The bright blue color of their cytoplasm after Giemsa's staining, and an early description by Koch, have led to the meronts being colloquially called "Koch's blue bodies".

The vectors, for all known species, are ixodid ticks. As with *Babesia*, the occurrence of sexual reproduction has only recently been confirmed. The process is less fully understood than in *Babesia*, but seems to be essentially similar. Kinetes, however, have not been seen to develop in cells of any organs of the tick other than salivary glands. Transmission from adult to larva via the eggs does not, therefore, occur, but *Theileria* may persist in ticks from nymph to adult stages. The life cycle of *Theileria* is outlined in Figure 8.7.

Pathogenesis

The theilerioses occurring in domestic cattle, sheep, and goats vary from mild to acute. They may often be fatal, febrile diseases with enlargement of the lymph nodes and spleen, and congestion of the lungs and meninges (the membranes surrounding the brain). Ulcers sometimes develop in the abomasum and intestine. Anemia may occur if the infection is heavy. Hemoglobinuria is much less common than in babesiosis, though it may occur in disease in cattle due to *T. annulata* and, briefly, in that due to *T. hirci* of sheep and goats. The most pathogenic species are *T. parva* (mortality rate up to 90%) and *T. annulata* in cattle, as well as *T. hirci* in sheep and goats. The species found in English and North American cattle is only slightly pathogenic, if at all; it was called *T. mutans* until recently, but is now thought to represent another, as yet undefined, species.

Diagnosis

Diagnosis of all these species depends on the finding of parasites in blood films (Fig. 8.5) and in smears of material obtained by lymph gland puncture, where exoerythrocytic meronts may be seen (Fig. 8.6).

GENUS *CYTAUXZOON* NEITZ AND THOMAS, 1948

Cytauxzoon has been recorded from three species of African wild ungulates (duiker, kudu, and eland) and from domestic cats in the United States. The erythrocytic forms are similar to those of *Theileria* but the exoerythrocytic forms are in the vascular endothelium rather than in lymphocytes. The *Cytauxzoon* of the domestic cat (*C. felis*) is reported to be pathogenic for that host. Infections in domestic cats are

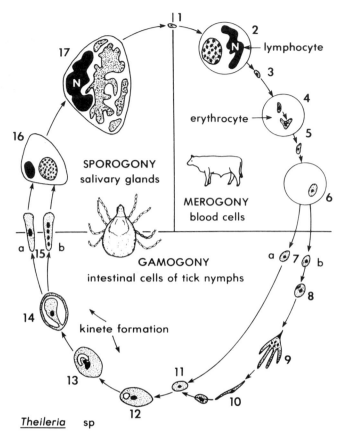

Figure 8.7 Diagram of life cycle of *Theileria* species. (1) Sporozoite in saliva of a feeding tick. (2) Meront ("Koch's blue body") inside lymphocyte (N = lymphocyte nucleus). (3) Merozoite. (4–5) Binary fission inside erythrocyte. (6) Ovoid stage inside erythrocyte. (7a–b) Ovoid stages free in blood masses in tick gut. (8–10) Formation of microgamonts (9) and microgametes (10). (11) Macrogamete. (12) Zygote. (13–15) Formation of motile kinete from ovoid stationary zygote. (15b) In *T. parva* division of the nucleus may start in kinetes before they leave the intestinal cells of the tick. (16) Kinete that has entered a salivary gland cell and started asexual reproduction by growth and nuclear division. (17) Enlargement of host cell and its nucleus (N); inside the giant cell thousands of sporozoites are formed (from H. Mehlhorn & E. Schein 1984).

accidental; the reservoir is some wild animal, probably the American bobcat. The exoerythrocytic forms are in phagocytic cells in the vascular endothelium in many organs of the body.

Treatment

The compound called menoctone (a hydroxyalkylated naphthoquinone) and some related compounds, as well as halofuginone, have good antitheilerial activity. It is possible that they may provide effective therapy for the theileriosis. Infection can be prevented only by (a) attacking the tick vector by dipping and spraying the animals, (b) good fencing to keep out wild ungulates which may serve as reservoirs of infection (e.g. buffalo, in the case of *T. parva* in East Africa), and (c) quarantine measures directed against outbreaks. Animals recovering from infection are resistant to reinfection, though with all species except *T. parva* this appears to be dependent upon the persistence of small numbers of viable parasites in the host, and hence such animals are potential reservoirs of infection. Vaccination with live, avirulent strains, or with virulent strains followed by treatment with long-acting tetracyclines, has been used with some success to prevent acute *T. annulata* infection.

FAMILY DACTYLOSOMIDAE

This family contains piroplasms of cold-blooded vertebrates, of which two genera have been described. Little is known of the pathogenicity and life cycle of these genera. They do not appear to harm their hosts, and only stages in erythrocytes have been seen. Nothing is known of their transmission. Meronts are seen in the red blood cells, and also nondividing forms which are thought (with no direct supporting evidence) to be gametocytes. The two genera which have been described are *Dactylosoma* Labbé, 1894 (recorded from reptiles, amphibia, and fish in various, scattered parts of all the continents except Australasia), and *Babesiosoma* Jakowska and Nigrelli, 1956 (so far recorded only from amphibia in North America and fish in Africa). *Dactylosoma* produces from 4 to 16 merozoites, arranged in a fan-like pattern within the erythrocytes, while *Babesiosoma* has meronts closely resembling those of some species of *Babesia*, with only four merozoites.

An organism named *Anthemosoma garnhami*, discovered in the erythrocytes of the African spiny mouse (*Acomys percivali*), has been provisionally placed in this family. Levine thinks it should be given a family of its own, the Anthemosomatidae.

Several other "genera" were at one time thought to belong to this class, or close to it. They are now known to be not protozoa but bacteria (*Bartonella*), rickettsiae (*Anaplasma, Eperythrozoon, Haemobartonella, Aegyptianella*), or viruses (*Pirhemocyton*).

FURTHER READING

Barnett, S. F. 1977. *Theileria*. In *Parasitic protozoa*, Vol. 4, J. P. Kreier (ed.), 77–114. New York: Academic Press.

Irvin, A. D. & M. P. Cunningham 1981. East coast fever. In *Diseases of cattle in the tropics*, M. Ristic & I. McIntyre (eds), 393–410. The Hague: Martinus Nijhoff.

Levine, N. D. 1985. *Veterinary protozoology*. Ames, Iowa: Iowa State University Press.

Mahoney, D. F. 1977. *Babesia* of domestic animals. In *Parasitic protozoa*, Vol. 4, J. P. Kreier (ed.), 1–52. New York: Academic Press.

Mehlhorn, H. & E. Schein 1984. The piroplasms: life cycle and sexual stages. *Advances in Parasitology* **23**, 37–103.

Ristic, M., P. Ambroise-Thomas & J. P. Kreier 1984. *Malaria and babesiosis*. Dordrecht, The Netherlands: Martinus Nijhoff.

Ristic, M. & G. E. Lewis, Jr. 1977. *Babesia* in man and wild laboratory adapted mammals. In *Parasitic protozoa*, Vol. 4, J. P. Kreier (ed.), 53–76. New York: Academic Press.

Uilenberg, C. 1981. *Theileria* infections other than east coast fever. In *Diseases of cattle in the tropics*, M. Ristic & I. McIntyre (eds), 411–42. The Hague: Martinus Nijhoff.

Zwart, D. & D. W. Brocklesby 1979. Babesiosis, non-specific resistance, immunological factors and pathogenesis. *Advances in Parasitology* **17**, 49–113.

CHAPTER NINE

Myxozoa, Microspora, and Ascetospora

For a long time classified as part of the "Sporozoa", the three groups Myxozoa, Microspora, and Ascetospora, are now each accorded independence as phyla. All these organisms produce resistant, thick-walled spores often containing one or more long polar filaments which are extruded when development in a new host occurs. The spores also contain one or more "sporoplasms" (germinative cells). All are symbiotic.

PHYLUM MYXOZOA

The spore is multicellular and of complicated structure. These organisms infect annelids and sipunculids, arthropods and cold-blooded vertebrates. Some may infect two hosts alternately in their life cycles.

ORDER BIVALVULIDA

In this order, one or (usually) more polar filaments are present, coiled within special polar capsules (Fig. 9.1). The spore wall is composed of two or more distinct valves, rather like the shell of bivalve molluscs (e.g. mussels – *Mytilus* spp.), the line where they join being called the sutural line. All are symbiotic in cold-blooded vertebrates, almost exclusively fish, and have a world-wide distribution. In this book a classification based on the number of valves forming the shell has been followed, with the position of the polar capsules determining division into suborders (Table 9.1).

Many Myxozoa, particularly those which develop in the viscera, produce serious disease in their hosts. Such diseases may be fatal, like

203

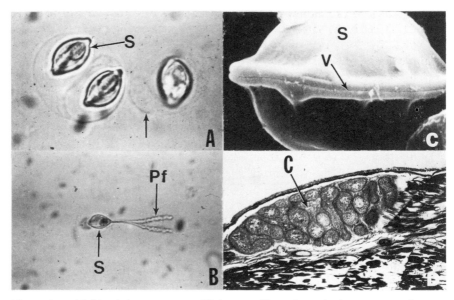

Figure 9.1 (a) *Myxobolus* sp. spores (S) in sporoblasts (arrow); there are usually two spores in each sporoblast; (b) light micrograph of a *Myxobolus*: spore (S) with polar filaments (Pf) extended; (c) scanning electron micrograph of *Myxobolus* spore (S) showing the junction of the two valves (V); (d) *Myxobolus* cysts (C) in tissue; these cysts may be large enough to be seen with the naked eye (micrographs kindly provided by Dr L. G. Mitchell, Department of Zoology, Iowa State University, Ames, Iowa).

Table 9.1 Classification of the Myxozoa.

Class Myxosporea Bütschli, 1881

Order 1. Bivalvulida (having 2 shell valves)

Suborder 1. Bipolarina (polar capsules at opposite poles of the spore)

2. Eurysporina (polar capsules at anterior pole and lying in a plane perpendicular to that of the sutural line)

3. Platysporina (polar capsules at anterior pole and lying in the plane of the sutural line)

Order 2. Multivalvulida (having more than 2 shell valves)

"twist" disease of salmon and trout, which is due to a myxosporidan *Myxobolus cerebralis* (also known as *Myxosoma cerebralis*). This species develops in the cartilage and perichondrium, including that of the skull, causing distension and consequent pressure on the brain and spinal cord. Many species infect the skin and muscles of their hosts, and when the latter are food fish the damage done, and the unprepossessing appearance of infected fish (the organisms' cysts are visible to the naked eye), may be of considerable economic importance; "tapioca" disease of salmon in the Pacific, caused by *Henneguya salminicola* in the muscles, is an example. It also seems probable that heavily infected fish, particularly when organs such as the liver are involved, may grow more slowly (though there is no direct evidence of this). In Lake Victoria, East Africa, virtually all the food fish are infected; thus, even quite a small reduction in the growth rate of infected individuals would have a considerable effect on the biomass of these fish, and hence on the protein available for food around the lake shore: an area in which protein is already a scarce commodity.

Morphology and life cycle

The life cycle of the group is imperfectly known, though the general pattern seems clear. After spores enter a vertebrate host it is thought that the polar filaments are everted to serve as an "anchor" (Fig. 9.1). The sporoplasms then emerge, penetrate the gut wall and somehow (possibly via the bloodstream) reach the appropriate host organ or tissue. Here they become trophozoites and increase in size, with membranous divisions developing, to form large (often macroscopic) cysts with a definite limiting membrane. In the forms which inhabit body cavities (coelozoic species), such as gall or urinary bladders or kidney tubules, the cysts lie entirely free. In the forms which inhabit tissues (histozoic species), the cysts are embedded in the tissue (muscle, skin, cartilage, liver, spleen, kidney, etc.) but are apparently extracellular; sometimes they appear to be in or alongside small blood vessels.

Within the cysts, spores develop. Certain cells (sporonts) become differentiated from the syncytial mass. The nucleus of each sporont divides several times to form a sporoblast containing, probably, six nuclei. In most genera, two sporoblasts develop from each sporont; they remain within a common membrane to form what is known as a pansporoblast. The cytoplasm then divides into four uninucleate cells and one which is binucleate. Two of the uninucleate cells form the shell halves, and two form the polar capsules. The binucleate cell becomes the sporoplasm and, at some stage of its development, the two nuclei apparently fuse. This is thought to represent an autogamous sexual process.

205

When cysts develop within the kidney tubules, urinary bladder, and gall bladder, or in the skin or muscle of the body wall or gills, it is easy to see how the spores could be liberated by the bursting of the cyst. The spores of species whose cysts develop in tissues such as cartilage or deep muscle, or organs such as the spleen, are presumably dependent upon the death of the host for dissemination.

The spores produced within the cysts developing in vertebrates are very variable in shape between the different myxozoan genera, but all conform to a basic pattern, which can be described with reference to one of the genera having less specialized spores, *Myxobolus*, a member of the suborder Platysporina (Table 9.1). Each shell valve is a flattened convex oval, and the two halves are joined in a sutural ridge at the margin of the spore. Within, at the anterior end, are the two polar capsules, each containing a coiled polar filament. Behind the polar capsules is the sporoplasm, a small mass of cytoplasm containing two nuclei which later fuse. The spores of different species vary widely in size, usually between about 10 and 20 μm in length. The spores of some genera (e.g. *Henneguya*) have long posterior processes, possibly as an aid in floating, while others (e.g. *Ceratomyxa*) are elongated laterally, perhaps for the same reason.

The means of transmission from vertebrate to vertebrate has long been obscure; in 1984 Wolfe & Markiw reported that *Myxobolus* (= *Myxosoma*) *cerebralis* (of trout) underwent part of its life cycle in intermediate, vector hosts – aquatic annelids of the family Tubificiidae. The worms were said to ingest spores of the type described above, which hatch in the gut and initiate an infection with organisms known for a long time as *Triactinomyxon*, and classified as a so-called second class of Myxozoa, the Actinosporea. Within the worm's gut, this stage of the parasite develops as sporocysts containing eight three-valved, anchor-shaped, actinosporean-type spores, each with three polar capsules and containing numerous sporoplasms. These spores are then ingested by the fish, either after liberation in the tubificiids' feces or while still contained within the worms, on which the fish avidly feed. Infection via the gills is also possible. The spores then hatch and the contained sporoplasms migrate (possibly via the bloodstream) to the appropriate location within the fish to initiate the myxosporean phase of the life cycle described above.

This life cycle has so far been claimed for only one species, and other workers have failed to confirm it. If a similar cycle occurs in all species of Myxosporea and Actinosporea, the latter class would cease to exist as a distinct taxon. As the generic name *Triactinomyxon* was proposed later than *Myxobolus* (and *Myxosoma*), the correct name for the parasite causing whirling disease in trout would remain *M. cerebralis*.

Pathogenesis

The pathological processes involved in myxosporidan infections are little known. Many of those which form visible cysts on the integument and gills of the host probably do it little damage, and the same may well be true of the coelozoic species. Others, however, produce recognizable disease. *Kudoa thyrsites* infects the musculature of the Australian and African barracuda (*Thyrsites atum*) and produces liquefaction of the surrounding muscle fibers, presumably resulting in muscle weakness. *Myxobolus cerebralis*, in the cartilage and perichondrium of American salmonid fish, produces distortion of the skeleton and hence of the body of the fish, presumably by some local cytotoxic effect. *Sphaerospora tincae*, a histozoic parasite of the European tench (*Tinca tinca*), causes distension of the fishes' abdomen which may eventually lead to death by rupture of the abdominal wall.

Myxosporidan infections are diagnosed by identifying the characteristic spores in fresh or stained smears, or in sections, from infected organs. Nothing is known regarding treatment. With the development of intensive "fish farming" myxosporidioses may become of increasing economic importance. Prevention of infection by those species (if any) having an annelid vector could be achieved by rearing fish in tanks from which the vector worms were excluded, and by filtering water entering the tanks to remove contaminating "actinosporean" spores.

PHYLUM MICROSPORA
CLASS MICROSPOREA

Two orders are recognized within this class, the Minisporida and the Microsporida; only the latter are considered in this book. Members of this phylum are differentiated from the Myxozoa by possessing unicellular spores. Microsporea are symbionts mainly of arthropods and fish, and more rarely of amphibia. Two species infect mammals; one of these has been rarely recorded from man, and there is a single record of a third species infecting man (see below). Microsporea are found throughout the world, and some are pathogenic, causing diseases in, for example, economically important insects such as the silk-worm (pébrine disease, due to *Nosema bombycis*) and honey-bees (nosema disease, due to *N. apis*). Colonies of anopheline mosquitoes, raised in laboratories for experimental work on malaria, sometimes suffer from epizootic disease of the larvae caused by *Thelohania legeri* or other Microsporida.

Morphology and life cycle

The spores, which are the infective stages, are small and simpler in structure than those of the Myxozoa (Fig. 9.2). They are spherical, oval or cylindrical and vary from 5 µm or fewer to 10 µm in length. The

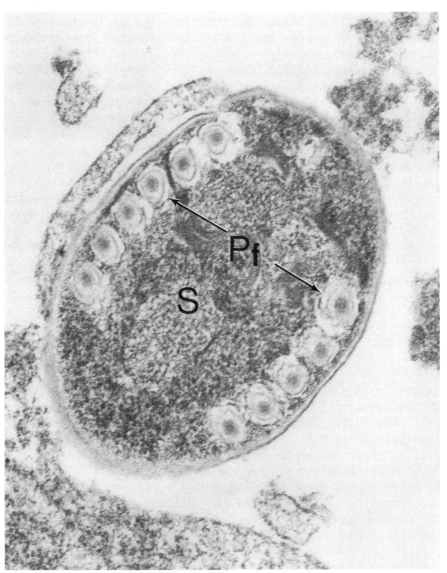

Figure 9.2 Electron micrograph of rabbit *Encephalitozoon cuniculi* grown in rabbit choroid plexus cells in culture. The micrograph shows a cross section of a spore (S) with polar filaments (Pf) (provided by Professor J. A. Shadduck, College of Veterinary Medicine, University of Illinois, Urbana).

wall, apparently not composed of separate valves, is chitinous. Within it lies a single, probably uninucleate, sporoplasm and, in front of this, a body called the polaroplast which is believed to cause extrusion of the polar filament by swelling. The single polar filament is coiled around both sporoplasm and polaroplast, and is not contained within a separate polar capsule (Fig. 9.3).

When the spore is swallowed, the polar filament is extruded in the lumen of the host's gut and the sporoplasm migrates along the hollow filament to emerge as a small ameba. In some species, at least, the filament penetrates a gut cell and so the emerging ameba finds itself already within a host cell: in other species, the ameba probably penetrates the gut wall after emergence. It eventually (somehow) reaches its chosen location in the body, where it enters a cell, grows and divides repeatedly (by a process considered to be merogony) until the now very enlarged cell becomes filled wth unicellular sporonts. These aggregations are often incorrectly called "cysts"; since their only limiting membrane is that of the host cell they should be referred to as pseudocysts. Each sporont then gives rise to one or more spores, the number differing in different genera. Figure 9.4 outlines the life cycle of *Encephalitozoon cuniculi*.

When the pseudocysts occur in deep tissues of a vertebrate host, the release of the spores is presumably dependent upon the host's death; in other instances, however, they may be liberated directly from a rupturing pseudocyst in the skin, or into various body cavities which are connected with the outside world (e.g. kidney tubules of vertebrates). Release from infected insects is probably usually depen-

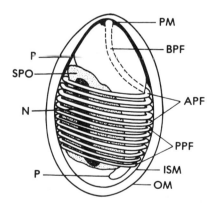

Figure 9.3 Diagram of the structure of a microsporidan spore, greatly enlarged; anterior polar filament (APF); basal portion of the filament (BPF); inner surface membrane (ISM); nucleus (N); outer membrane (OM); polaroplast (P); polar mass and polaroplast membrane (PM); posterior polar filament (PPF), sporoplasm (SPO). (From R. R. Kudo 1966, courtesy of C. C. Thomas, Springfield, Illinois).

209

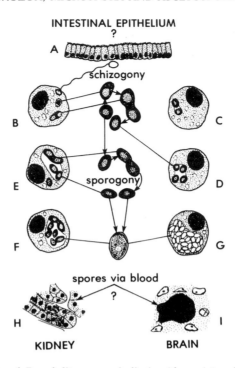

INTESTINAL EPITHELIUM

Figure 9.4 Life cycle of *Encephalitozoon cuniculi*. A wide variety of mammals can be infected by ingestion of spores. The hatching of spores and the migration from the intestine (a) to the viscera have not been observed. Asexual division by binary fission ("schizogony") takes place in peritoneal macrophages (b, c, d), and results in the development of a parasitophorous vacuole. Sporonts with thickened walls give rise to pairs of sporoblasts (e, f), which develop into spores. Repeated sporogony results in macrophages packed with spores (g). In chronic infections, development localizes in cells of the viscera, especially in the kidney (h) and brain (i). Spores are passed out with the urine (From E. U. Canning 1977).

dent on the death of the host, though some species are known to be transmitted via the eggs to the larvae (e.g. *N. bombycis* of the silkworm).

Although all Microsporida have unicellular spores, as described, in one genus with a single species (*Telomyxa glugeiformis*) the spore are stuck together in pairs by the posterior ends so that, at first sight, each spore seems to have two sporoplasms and two polar filaments.

Pathogenesis

Little is known about the pathogenesis of microsporidan infections, though many (especially among insects) are fatal. Infected cells usually increase enormously in size, as do their nuclei, and are presumably rendered unable to perform their natural function. Thus, if an organ is heavily infected it may become virtually useless to the host and this, if the organ is a sufficiently important one, may lead to death of the host. For example, *Pleistophora myotrophica* of the toad (*Bufo bufo*) infects the skeletal muscles, which atrophy. If heavily infected, the toad becomes so weak that it dies from exhaustion, being presumably unable to feed. Sometimes these protozoa exert a harmful effect on organs other than those which they infect. For instance *N. apis* infects only the gut cells of the honey-bee, but the ovaries of infected queens degenerate – again, possibly as a result of malnutrition.

Microsporidan infections are diagnosed by recognizing the spores in sections of smears of infected organs. Little is known about their treatment or prevention, though the antibiotic Fumagillin is used to control *N. apis* of bees.

Encephalitozoon cuniculi is well known as a parasite of many species of rodents, insectivores, rabbits, carnivores, and primates. It inhabits various cells, but preferentially invades peritoneal macrophages in which binary fission occurs. The central nervous system and kidney tubule cells are invaded later in the chronic phase, and aggregations of small oval spores (about 2.5×1.5 μm) develop therein (Fig. 9.5). Transmission occurs by ingestion of food or water contaminated with urine containing spores, and carnivores may well become infected from eating parasitized prey. One human infection with *E. cuniculi* was reported in 1959, involving the cerebrospinal fluid and urine of a Japanese boy who, although seriously ill with encephalitis, subsequently recovered; another human infection, with what was probably the same species, was reported in 1973, the parasites being in the cornea of a boy in Ceylon. There have been a few other records from primates, all in monkeys.

Thelohania apodemi has been reported from the brains of field mice (*Apodemus sylvaticus*) in France, and there has been a single report of disseminated infection with a species possibly belonging to the genus *Nosema* (named *N. connori*) in an immunologically incompetent child aged 4 months, who died from *Pneumocystis carinii* pneumonia. This case was possibly an "adventitious" infection, i.e. establishment of infection in a normally insusceptible but immunologically defective host. Such cases may become more common with the spread of HIV virus and the associated AIDS. (Most, if not all, species of *Nosema* parasitize invertebrates; at one time *E. cuniculi* was wrongly thought to be a species of *Nosema*.)

No drug treatment is known for any microsporan infection of mammals.

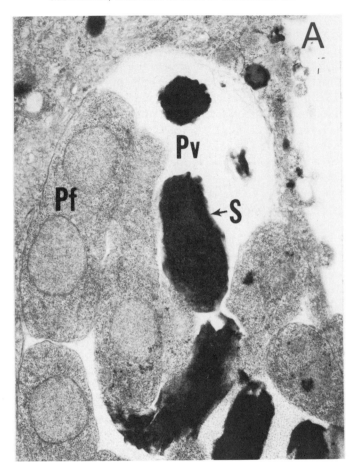

PHYLUM ASCETOSPORA

These organisms produce multicellular spores containing one or more sporoplasms but lacking polar capsules and polar filaments.

The Ascetospora are parasites of invertebrates and fishes. In the connective tissue or hemocoel of the host they are extracellular and undergo a type of merogony resulting in the formation of binucleate cells, which, by a process called "sporogony" but apparently not homologous with sporogony of the Telosporea since sex is not involved, differentiate into spores. The spore is the infective phase: when one is swallowed by a new host, a small, ameboid organism emerges from it and migrates to its site of development. Several genera

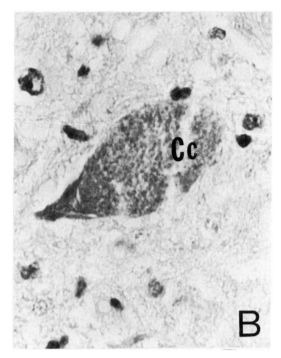

Figure 9.5 (a) Electron micrograph of a canine embryo fibroblast cell in culture which is infected with canine *Encephalitozoon cuniculi*. Proliferative forms (Pf) undergoing binary fission are present at the periphery of a parasitophorous vacuole (Pv). The dark, underexposed structures towards the center of the vacuole are tangential sections of spores (S). (b) Photomicrograph of a Giemsa-stained section of a cerebral cortex brain cell (Cc) filled with spores and proliferative forms of *E. cuniculi*; the section is from a dog with canine encephalitozoonosis (micrographs provided by Professor J. A. Shadduck, College of Veterinary Medicine, University of Illinois, Urbana).

and species have been described. *Minchinia nelsoni* infects the American oyster (*Ostrea virginica*) and has devastated oyster populations of the Delaware and Chesapeake bays and other regions of the eastern United States coast. The disease seems to be spreading to other parts of the world (1987).

FURTHER READING

Canning, E. U. 1977. Microsporida. In *Parasitic protozoa*, Vol. 4; J. P. Kreier (ed.), 155–97. New York: Academic Press.

Cheng, T. C. 1973. *General parasitology*. New York: Academic Press.

Mitchell, L. G. 1977. Myxosporida. In *Parasitic protozoa*, Vol. 4; J. P. Kreier (ed.), 115–54. New York: Academic Press.

Wilson, J. M. 1979. The biology of *Encephalitozoon cuniculi*. *Medical Biology* **57**, 84–101.
Wolf, K. A. & M. E. Markiw 1984. Biology contravenes taxonomy in the Myxozoa: new discoveries show alternation of invertebrate and vertebrate hosts. *Science* **225**, 1449–52.

CHAPTER TEN

Symbiotic ciliates

———————

The ciliates constitute a class apart, indeed a phylum apart, among the protozoa. They are a vast group and a vast subject, and will be dealt with here only very briefly. A detailed taxonomic and morphological account of the ciliates, symbiotic and free-living, can be obtained from the books listed at the end of this chapter. The classification used here is that published in 1980 by the committee on systematics and evolution of the Society of Protozoologists, though more recent variants have been proposed. The great majority of ciliates are free-living, symbiosis having apparently arisen independently among various groups. Indeed, within the single genus *Tetrahymena* (subclass Hymenostomatia, order Hymenostomatida), all gradations of adaptations to symbiosis occur. There are species which are entirely free-living, those which can live equally well both free or as symbionts, species which are almost entirely symbiotic with only occasional periods of "free" existence during their life cycles (facultative symbionts), and species which are entirely symbiotic (obligate symbionts).

A few orders and suborders consisting of symbiotic forms have evolved among the subclass Hymenostomatia: these include the Astomatida (endosymbiotic in oligochaete annelids) and the Thigmotrichida (endo- and ectosymbiotic in or on bivalve molluscs). Some members of the suborder Ophryoglenina are endo- or ectosymbionts. *Ichthyophthirius*, a common and sometimes harmful ectosymbiont of fish, including aquarium fish, is the best known of these. In the subclass Gymnostomatia, the order Entodiniomorphida is a very specialized group inhabiting the stomachs of herbivores. Many of the sessile forms in the subclass Suctoria are attached to various aquatic animals, and may sometimes be injurious to their hosts. The only ciliate causing disease of man is *Balantidium coli* (subclass Vestibuliferia, order Trichostomatida).

All ciliates are characterized by the possession of cilia during part or all of their life cycle. The basal bodies of these cilia, together with an associated complex of fibrils, are present throughout the life cycles of all species. Ciliates almost always have nuclei of two types, a macro-nucleus and one or more micronuclei, within each individual. The macronucleus is concerned with the day-to-day somatic functioning of the organism, the micronucleus with sexuality. Sexual reproduction takes the form of the unique process known as conjugation which, basically, occurs as follows. Two ciliates pair and become united by the temporary dissolution of the contiguous parts of their pellicles; their macronuclei disintegrate and their micronuclei divide twice, one of these divisions being meiotic; three of the resulting micronuclei degenerate, while the fourth divides once more to produce two gametic nuclei. One of the gametic nuclei (the "male") from each member of the conjugating pair migrates into the other member of the pair, where it fuses with the stationary ("female") nucleus. The zygotic nucleus then divides twice and the pair of ciliates separates; both individuals then divide, each daughter receiving two nuclei, one of which becomes polyploid and forms the new macronucleus while the other constitutes the new (diploid) micronucleus.

GENUS *BALANTIDIUM* CLAPARÈDE AND LACHMANN, 1858

This genus contains *B. coli*, the only species of ciliates which infects man. About 50 species of *Balantidium* have been described, mainly from primates and amphibia, but some are probably synonyms; other animals which have been recorded as hosts of *Balantidium* include pigs, sheep, guinea-pigs, camels, opossums, ostriches, fish, cockroaches and other insects, flatworms, and coelenterates.

Specimens of *Balantidium* for study may be obtained from pig feces or the rectal contents of frogs (killed with ether or chloroform). Another ciliate (*Nyctotherus cordiformis*) may also be found in the rectum of frogs, but can be distinguished from *Balantidium* by the lateral position of its "mouth" (peristome). Another, taxonomically entirely distinct, organism occurs in the rectum of frogs: the flagellate *Opalina*; this can be easily distinguished from *Balantidium* as it is much larger.

Balantidium coli inhabits the large intestine (cecum and colon) of pigs, man, apes, and monkeys; it has also been recorded occasionally from dogs, rats, sheep, and cattle. The pig is the commonest and probably the main natural host. *Balantidium coli* has been recorded from all parts of the world where pigs occur. In man (and other primates) *B. coli* causes a disease called balantidiosis or balantidial dysentery. In pigs it is generally nonpathogenic. The ciliate is relatively large and ovoid (Fig. 10.1), and is usually about 60–70 × 40–60 μm (though larger and smaller individuals have been reported). The entire surface is covered

216

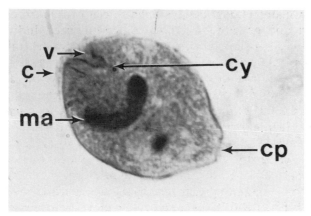

Figure 10.1 *Balantidium coli*; trophozoite showing various structures: cilia (c); cytopyge (cp); cytostome (cy); macronucleus (ma), vestibule (v).

with cilia arranged in longitudinal, slightly spiral rows. At the front is a deep groove or vestibule, lined with slightly longer cilia, which leads to the mouth or cytostome. The animal feeds by drawing a current of water down the vestibule by means of the cilia lining the latter. Any food particles carried in this current are ingested through the cytostome in small food vacuoles, which then circulate along a defined path through the ciliate's body while their contents are being digested. Eventually any undigested remnants are expelled through a permanent pore or cytopyge near the organism's hind end. The parasite has two contractile vacuoles, a macronucleus and micronucleus.

Balantidium coli reproduces asexually by transverse (i.e. homothetogenic) fission, and sexually by conjugation followed by binary fission. The organism encysts in the lumen of the host's large intestine, and the cysts, passed out in the feces, are the transmissive stage. They can survive outside the host for some time, presumably hatching in the large intestine of a new host after being swallowed. The cysts are large and spherical (Fig. 10.2) and are about 50–60 µm in diameter; the contained individual organism often does not entirely fill the cyst. In young cysts, the cilia and contractile vacuoles of the parasite may be seen functioning, but eventually the encysted organism becomes quiescent. No nuclear or cytoplasmic division occurs in the cyst.

As noted earlier, the pig seems to be the natural host of *B. coli*, and man is an accidental host who usually becomes infected from a pig by swallowing cysts passed out in the latter's feces. Consequently, most human infections are in country-dwellers and farmworkers. Human infections are rare; by contrast most pigs are infected.

In man, ciliates not only inhabit the lumen of the large intestine but also sometimes invade the intestinal wall, penetrating the mucosa and

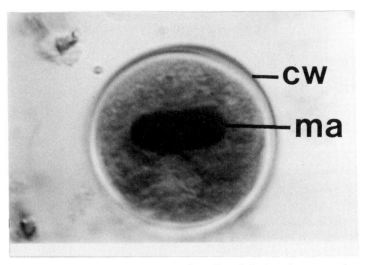

Figure 10.2 *Balantidium coli*; cyst showing the cell wall (cw) and macronucleus (ma).

submucosa and producing lesions like those caused by invasive *E. histolytica*. The irritation produced by the ciliates leads to diarrhea, and as the lesions progress blood vessels become eroded and bloody dysentery results. If untreated, the condition may progress to potentially lethal perforation of the large intestine. In sections of infected areas of intestine, the ciliates are easily recognized because of their large size and large macronuclei; they are often found in groups known as "nests" or (in Latin) "nidi" (singular, "nidus") (Fig. 10.3). In the pig the ciliate is normally noninvasive and, therefore, nonpathogenic. Apparently it cannot penetrate the intact intestinal mucosa, but if the latter is damaged by infection with bacteria such as *Salmonella*, or some other cause, invasion can occur resulting in lesions similar to those seen in man. Whether the ciliate also needs help to penetrate the human mucosa is not known; quite possibly it does.

Balantidium infection can be diagnosed easily by finding the ciliates, free or encysted, in fecal specimens. The organisms will grow in the culture media ordinarily used for intestinal amebae, but recourse to this aid is seldom necessary for diagnosis.

In man, the infection can be treated with various anti-*Entamoeba* drugs. The standard treatment is with metronidazole, although another nitroimidazole, nitrimidazine (nimorazole), appears to be even more effective. Treatment of pigs is unnecessary as the parasite is usually nonpathogenic to them. Prevention of infection in man is simply a matter of elementary sanitary precautions and the avoidance of a diet which includes pig feces.

218

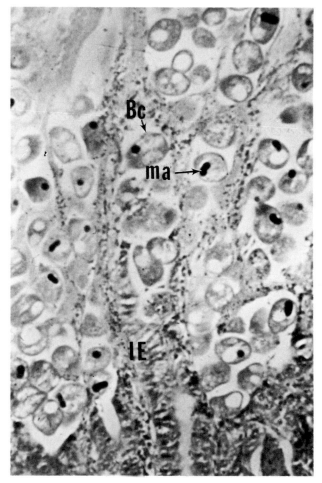

Figure 10.3 Photomicrograph of a "nest" of *Balantidium coli* (Bc) in the intestinal mucosa (IE). The macronuclei (ma) are visible when in the plane of section. There is much destruction of the mucosa and many organisms in the tissue.

THE RUMEN CILIATES

A bewildering array of ciliates inhabits the rumen and reticulum of ruminants; another, and equally bewildering group, lives in the cecum and colon of equids (Table 10.1). Many of these forms have a complicated structure with skeletal polysaccharide plates in the pellicle and a very reduced ciliature (Fig. 10.4). As far as is known, none of these forms produces cysts. The rumen-dwellers are killed when they pass from the rumen to the host's abomasum, and transmission occurs

219

Table 10.1 Ciliates living in the rumen and reticulum of ruminant herbivores, and in the cecum and colon of equids.

Hosts	Subclass	Order	Suborder	Family	Genera
Ruminants	Holotrichia	Gymnostomatida	Rhabdophorina	Buetschliidae	*Buetschlia*
		Trichostomatida	—	Isotrichidae	*Isotricha,*
	Spirotrichia	Entodiniomorphida	—	Ophryoscolecidae	*Dasytricha* *Ophryoscolex, Entodinium, Diplodinium,* and 15 other genera
Equids	Holotrichia	Gymnostomatida	Rhabdophorina	Buetschliidae	13 genera (not *Buetschlia*)
		Trichostomatida	—	Blepharocorythidae	*Blepharocorys, Charonina, Ochoterenaia*
				Paraisotrichidae	*Paraisotricha*
	Suctoria	Suctorida	—	Acinetidae	*Allantosoma*
	Spirotrichia	Entodiniomorphida	—	Cycloposthiidae	7 genera

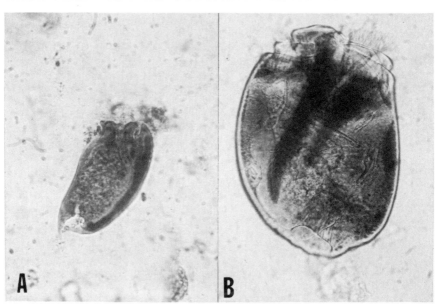

Figure 10.4 Rumen ciliates. (a) *Entodinium gibberosum*. Single oral ciliary zone at anterior end of cell. The macronucleus, with a notch in its anterior end, is visible on the right side of the body. (b) *Metadinium medium*. There are two ciliary zones, oral and dorsal, at the anterior end of the cell. Macronucleus, micronucleus, skeletal plates (two dark bars), and cytopyge are all visible. (c) *Dasytricha ruminantium*. Focused to show spiral rows of cilia which cover the entire surface of the cell. The macronucleus is also visible. (d) *Ophryoscolex caudatus*. The dorsal ciliary zone forms a band around three-fourths of the

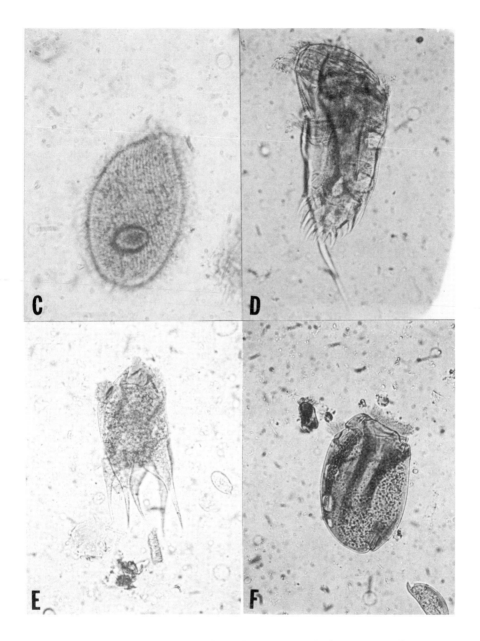

body, about one-third of the distance from the anterior end. Three circlets of secondary caudal spines are visible. (e) *Epidinium cattanei*. The dorsal ciliary zone is displaced anteriorly. Five large caudal spines are visible. (f) *Elytroplastron bubali*. There are two ciliary zones on the anterior end of the cell. Three (of four) skeletal plates, several contractile vacuoles, and cytopyge are all visible. (These photomicrographs were supplied by Dr B. A. Dehority, Ohio Agricultural Research and Development Center, Wooster, Ohio, USA).

when young animals feed on hay and grass contaminated with the saliva of older, infected animals; the saliva contains ciliates regurgitated during cud-chewing. The ciliates living in the large intestines of equids are presumably transmitted orally on food contaminated with feces containing the ciliates.

Little is known about the relationship of the ciliates of equines to their hosts, probably they are all harmless commensals. The host-relationships of the rumen ciliates have been much more studied; apparently none is harmful, most are harmless commensals, and a few are of definite value to their hosts. Unlike the symbiotic flagellates of termites, however, none of these forms is essential to its host. This subject will be dealt with only briefly here; more details can be obtained from the reviews listed under "Further Reading" at the end of this chapter.

Diplodinium and some other Spirotrichia can digest cellulose; there is doubt as to whether they do this by virtue of their own enzyme (cellulase) or that of symbiotic bacteria. These and many other genera also digest starch and soluble carbohydrates. Thus they may help the host in breaking down these substances, especially the cellulose (ruminants, like all mammals, do not produce a cellulase). However, the ruminants can live quite satisfactorily without the ciliates, cellulose digestion then being performed by the rumen bacteria alone. Other ways in which the ciliates may benefit the host are in storing carbohydrates and digesting them gradually, thus making available to the host a more regular supply of the volatile fatty acids into which the carbohydrates are converted (all rumen ciliates are obligate anerobes).

Although the ciliates convert some of the host's food into their own protein, this is not lost to the ruminant but becomes available to it when the ciliates die. This in fact benefits the host since the ciliate protein is of higher quality nutritionally than that provided by the plants on which it feeds or by the rumen bacteria. It has been estimated that at least 100 g of protein are provided daily by the rumen ciliates in an average ox – about one-fifth of the total daily protein requirement. Estimates of the numbers of ciliates present in the rumen vary widely, but they are all immense: values range from 100 000 to a million per cm^3, which, in an ox with 100 kg of rumen contents would amount to between 10 000 and 100 000 million (10^{10}–10^{11}) ciliates (2.5–25 times the human population of the world in 1975). Rumen and intestinal ciliates occur in all parts of the world wherever cattle, sheep, goats, camels, reindeer, water buffalo, elk, antelope, horses, donkeys or zebra occur.

FURTHER READING

Coleman, G. S. 1980. Rumen ciliate protozoa. *Advances in Parasitology* **18**, 121–73.

Corliss, J. O. 1979 *The ciliated protozoa*, 2nd edn. Elmsford, New York: Pergamon Press.

Hoffman, G. L. 1978. Ciliates of freshwater fishes. In *Parasitic protozoa*, Vol. 2, J. P. Kreier (ed), 584–632. New York: Academic Press.

Hungate, R. E. 1978. The rumen protozoa. In *Parasitic protozoa*, Vol. 2, J. P. Kreier (ed.), 655–96. New York: Academic Press.

Nisbet, B. N. 1984. *Nutrition and feeding strategies in protozoa*. London and Canberra: Croom Helm.

Zaman, V. 1978. *Balantidium coli*. In *Parasitic protozoa*, Vol. 2, J. P. Kreier (ed.), 633–54. New York: Academic Press.

CHAPTER ELEVEN

Techniques

INTESTINAL PROTOZOA

Symbiotic protozoa which inhabit the alimentary canal of their hosts may be obtained for study either post-mortem or by the collection of fecal specimens. Usually the latter contain only encysted forms, unless the stool is diarrheic, when trophozoites of amebae, flagellates and ciliates may be found. Many free-living protozoa (coprozoic forms) multiply rapidly in old, moist feces, and care has to be taken not to confuse them with symbiotic forms in samples which are not fresh. Specimens which cannot be examined immediately should be refrigerated at 4 °C if possible, and kept in closed vessels. Specimens may also be preserved by emulsification in 10% formol in 0.9% saline solution (formol–saline). "Formol" or "formalin" is a 40% aqueous solution of the gas formaldehyde (HCHO), thus a 10% formol solution contains 4% formaldehyde. Fecal samples may be examined either directly or after concentration.

Direct examination of fecal specimens

For direct examination, a small portion of the specimen (about the size of a large pin's head) is taken on a matchstick or swabstick and thoroughly dispersed in a drop of fluid on a microscope slide, covered with a coverslip and examined. If it is thought that trophozoites may be present, one preparation should be made in 0.9% saline; two other similar preparations should always be made, whether the presence of trophozoites is suspected or not, one in 1% aqueous eosin solution and one in double-strength Lugol's iodine solution (4% potassium iodide plus 2% iodine in distilled water). Practice is needed in making these preparations – if they are too thick it will be impossible to examine them microscopically, and if they are too thin, protozoa may not be seen unless present in very large numbers.

The saline preparation (if made) should be examined microscopically at a total magnification of about × 100, using phase-contrast illumination if possible, or with the intensity of illumination reduced by partially closing the substage iris diaphragm; any motile organisms seen can be examined further at a magnification of about × 400. The eosin preparation should be examined similarly. If it is correctly made (without an excess of eosin), living protozoan cysts (and helminth eggs) will appear as small unstained (white) objects against the pink background of eosin solution and the red-stained debris. They can thus be readily detected, and subjected to further examination at a magnification of × 400 and (using an oil-immersion objective) at × 500–× 1000, if necessary. Usually little detail can be seen in these unstained specimens, except for the chromatoid bodies which may be seen in cysts of *Entamoeba* spp. These should be searched for carefully and, if present, their shape noted. The shape and size of any cysts should also be noted.

If cysts have been seen during this preliminary examination of the eosin preparation, that suspended in iodine should be examined. Here the cysts will be less obvious, as almost everything is stained yellow-brown by the iodine, and it is better to commence the examination at a higher magnification (× 400); the oil-immersion objective can then be used to study individual cysts in more detail. Iodine stains the nuclei of the cysts, and their number and structure should be noted; also, vacuoles containing glycogen, if present, will be conspicuous as their contents stain a deep golden-brown color. Chromatoid bodies are not clearly seen in cysts stained with iodine.

Concentration of fecal specimens

Scanty infections are easily missed if only direct examinations of fecal specimens are made. If possible, a method of concentrating the cysts (and helminth eggs) in a specimen should be used (these methods, however, kill trophozoites). A simple method of concentration uses flotation in a zinc sulfate solution. A small portion of feces (about 15 cm^3) is emulsified in a 33.1% aqueous solution of zinc sulfate (ZnSO$_4$.7H$_2$O), using a small pestle and mortar. The emulsion is placed in a glass cylinder measuring about 5 × 2 cm, which must be filled to the brim, and then a coverslip is placed over the mouth of the cylinder, touching the fluid. After 20–30 minutes, the coverslip is gently removed and placed fluid side down on a microscope slide. Any cysts present should have floated up in the zinc sulfate solution and have been removed with the film of liquid attached to the coverslip. If cysts are present, they can subsequently be stained by placing a drop of double-strength Lugol's iodine solution alongside the coverslip and drawing the solution beneath the latter by applying a piece of filter paper to the opposite edge of the coverslip.

A better method of concentration is the formol–ether technique, which requires the use of a centrifuge. A 1–2 g sample of feces is emulsified in 10 cm^3 of 10% formol–saline and strained through a wire sieve (16 mesh cm^{-1}) into a small centrifuge tube; formol–saline is added to bring the fluid level to about 2.5 cm of the top of the tube; about 3 cm^3 of ether are then added, the tube is shaken vigorously and then centrifuged, the speed being increased gradually to a maximum of 2000 r/min (700g) after 2 min and the centrifuge then switched off. When the tube has come to rest, the fatty debris at the interface of the formol–saline and the ether is loosened from the wall of the tube with a matchstick or swabstick, and the debris plus supernatant fluid is poured off and discarded. The deposit is then resuspended in the small drop of fluid remaining in the tube, and picked up in a Pasteur pipette. Half is added to a drop of Sargeaunt's stain (0.2 g malachite green dissolved in 3 cm^3 of 95% ethanol; 3 cm^3 acetic acid is added and the volume made up to 100 cm^3 with distilled water) on a microscope slide; this stain colors chromatoid bodies and nuclei dark green. The other half of the resuspended deposit is added to a drop of double-strength Lugol's iodine solution to reveal the presence of any glycogen vacuoles.

The oocysts of intestinal coccidia are also concentrated by these two methods. However, in order to identify species it is often necessary to keep the oocysts *in vitro* at room temperature for 2–3 days for sporulation to be completed, which cannot be done after formol–ether concentration as this kills the parasites. Zinc sulfate solution is also likely to be harmful. The simplest method to concentrate coccidian oocysts is to suspend the feces in a saturated aqueous solution of sodium chloride and centrifuge the suspension at 1500 r/min (about 400g) for 2 min. The surface layer of solution, containing the oocysts, is at once pipetted into three or four times its volume of water, which dilutes the saline sufficiently to allow the oocysts to sink (by gravity or after further centrifugation). The sedimented oocysts are finally resuspended in 2% aqueous potassium dichromate solution and kept in a shallow layer of this solution in a Petri dish at room temperature until sporulation has occurred (as judged by the microscopical examination of a drop of the suspension).

Permanent preparations of intestinal protozoa

Permanently stained preparations of intestinal protozoa may be made by spreading a thin film of fecal matter on to a coverslip with a matchstick or swabstick and fixing it for 20–30 min in Schaudinn's solution while still wet. Schaudinn's solution contains 60 cm^3 of a saturated solution of mercuric chloride in 0.9% saline, 30 cm^3 of ethanol and 10 cm^3 of acetic acid. The fixed smear is briefly washed in 70% ethanol containing a little (3–5%) double-strength Lugol's iodine solution, to remove the mercuric ions introduced with the fixative, and then washed for five minutes by immersion in 70% ethanol containing 3–5% of 5% aqueous sodium hyposulfite. The preparation is then briefly passed through fresh 70% ethanol, 90% ethanol and pure ethanol (five min each, to harden the parasites), and rehydrated by passing through 70%, 50%, and 30% ethanol to distilled water. The preparation is then placed in tungstophosphoric acid–hematoxylin stain for 12 hours or longer: overstaining is impossible. The preparation is briefly washed in water, dehydrated through 30%, 50%, 70%, 90%, and pure ethanol, and cleared in xylene (*taking care not to inhale the toxic vapor*) or, preferably, a nontoxic clearing agent such as cedarwood oil or clove oil. The coverslip is finally mounted with Canada balsam (film downwards) on a slide.

The tungstophosphoric acid–hematoxylin stain is prepared as follows. Dissolve 0.1 g hematoxylin by heating in a little distilled water; when cool, make up to 80 cm^3 with distilled water and then add 20 cm^3 of 10% aqueous tungstophosphoric acid (analytical reagent grade). This solution must be allowed to "ripen" for several months before use (the "ripening" can be accelerated by adding 10 cm^3 of 0.25% aqueous potassium permanganate).

Cultivation of intestinal protozoa

Some intestinal protozoa grow readily *in vitro* at 37 °C in simple media such as Dobell and Laidlaw's "HSre + S". This medium is prepared as follows. Sufficient sterile horse serum is poured into sterile cotton-plugged test-tubes so that when the tubes are held at an angle of about 40°, the serum runs 3–4 cm up the tube. While lying at this angle, the tubes are heated at 80 °C in a water bath for 60–70 min to coagulate the serum.

Fresh hen's eggs are cleaned and sterilized by washing in alcohol and a small hole is cut in the blunt end of the egg with sterile scissors; each egg is then inverted over a flask containing sterile Ringer's solution (formula in g l^{-1}: NaCl, 9.0; KCl, 0.42; CaCl$_2$, 0.24; HaHCO$_3$, 0.2; glucose, 1.0). By puncturing the narrow end of the egg, the albumen is allowed to run out of the hole into the flask (four eggs are used for each liter of saline). If not prepared aseptically, this solution must be sterilized by filtration.

Just enough albumen–Ringer solution is added to the test tube to cover the coagulated serum. The completed tubes should be incubated at 37 °C for 24–48 hours to check their sterility, and may be stored in a refrigerator for up to 1 week before use. Just before use, a small amount of rice starch (sterilized by heating to 160 °C for 1 hour) is added to each tube.

The tubes should be warmed to 37 °C before inoculation; a little fecal matter is then introduced by means of a bacteriological wire loop to the bottom of the overlay. The inoculated tubes are incubated at 37 °C and examined daily for the first few days. Each day new tubes are inoculated from the initial tubes. Established cultures need be subinoculated only every 2–4 days. Subinoculations are usually made with a Pasteur pipette: the base of the slope is scraped to dislodge any amebae, and then 0.5–1 cm^3 of the sediment at the bottom of the tube is transferred to a new tube.

If this medium is being used for diagnosis, it is advisable to subinoculate every 2–3 days even if no ameba is seen, since it may take a few days for reasonable numbers of trophozoites to develop. Most intestinal flagellates of mammals, *Balantidium coli*, and amebae of the *Entamoeba histolytica* group, grow readily in this medium; *Giardia* is one of the few intestinal flagellates which will not grow in it. Other, more complex, media are described in books listed at the end of this chapter.

Histological sections

Tissues (e.g. intestinal wall) infected with protozoa such as *E. histolytica*, *Balantidium coli*, or *Histomonas meleagridis* can be fixed, embedded, and sectioned by the usual histological procedures, and stained with hematoxylin and eosin (details of the methods can be obtained from standard histology textbooks listed at the end of this chapter); alternatively, the Giemsa method for sections can be used.

Availability of material for study of intestinal protozoa

Obtaining parasites of man for study may be difficult, but acceptable substitutes can often be found in domestic and laboratory animals.

Rhesus monkeys may harbour most of the common human intestinal parasites, but a more readily available substitute host is the pig. Pig feces will usually contain cysts of *E. suis* (which resemble those of *E. histolytica* except that they are uninucleate), *Balantidium coli* and *Iodamoeba buetschlii*. Species of *Giardia* and *Trichomonas* can be obtained from laboratory rodents, especially hamsters and guinea-pigs; *Spironucleus* may also be present. Intestinal coccidia of the genus *Eimeria* are easily obtainable from laboratory or domestic rabbits and chickens; young animals are more likely to be heavily infected. Isosporine coccidia are commonly found in wild passerine birds. The rectum of frogs usually contains a ciliate called *Nyctotherus*, and sometimes also a species of

Balantidium (not *Balantidium coli*) and *Opalina*. Examination of one's own feces may sometimes be rewarding – and surprising.

TISSUE PROTOZOA

Protozoa living in the blood and other tissues of their hosts can be obtained for study either after the death of the host or sometimes, depending on the tissue which they inhabit, during life by removing a small portion of the appropriate tissue. The latter procedure should be attempted only by qualified workers with access to suitable equipment and anesthetic and operating facilities, and care must be taken that it does not infringe the legislation of the country in which it is being done.

Blood films

Small quantities of blood can be simply obtained (subject to the above provisions) from small rodents by pricking the tail near its tip with a needle, or by cutting off its extreme tip with scissors. For larger mammals, pricking an ear vein is often satisfactory; with man, the finger tip may be pricked with a sterile needle (having sterilized the skin with alcohol and allowed it to dry). Having obtained a drop of blood, films may be prepared in two ways.

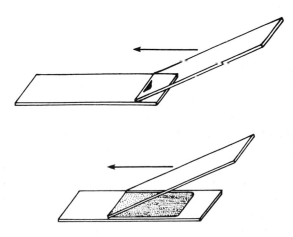

Figure 11.1 Diagram illustrating the preparation of a thin blood film (see text for details).

Thin blood films   The blood drop is placed near one end of a microscope slide; another slide is then placed with its narrow edge touching the drop and inclined so that the drop runs along the acute angle (about 30°) formed by the two slides; the inclined slide is then pushed along the horizontal one in the direction of the obtuse angle between the two slides, so that the blood drop is pulled (not pushed) along the horizontal slide (Fig. 11.1). The more acute the angle between the two slides, and the more slowly the "spreader" slide is moved, the thinner the film will be. The resulting film is then dried rapidly by waving it in the air (not by heating), and should be fixed and stained as soon as possible. If storage before staining is essential, the slide should be kept in a desiccator and preferably in a refrigerator (4 °C).

Thin blood films are fixed and stained as follows.

(1) Fix the film by covering it with methanol (= methyl alcohol) for half a minute or longer (fixation is almost instantaneous).
(2) Shake off the excess methanol (it does not matter if the film dries) and place the slide in a solution of Giemsa's stain (1 volume) in phosphate buffered water at pH 7.2 (9 volumes) for 50–60 min.
(3) Remove the slide from the stain and wash it by flooding *very briefly* (not longer than one second) under a tap.
(4) Place the slide in an upright position to drain and dry (do not use heat).

Not all brands of Giemsa's stain are equally satisfactory: three reliable brands are those supplied by Merck (obtainable from British Drug Houses), the National Aniline Division of Allied Chemical and Dye Corporation, and the Fisher Scientific Company (suppliers' addresses are given at the end of this chapter). The pH of the water used for diluting the stain may be controlled by buffering with 3.0 g l^{-1} Na$_2$HPO$_4$ and 0.6 g l^{-1} KH$_2$PO$_4$. It is best to place the slides in the stain face downwards to avoid deposition of scum, etc. This can be done by using specially designed trays or dishes, or by using watchglasses of diameter about 5 cm filled to the brim with stain, or Petri dishes containing two pieces of glass rod to support the slides.

Thin films result in the best morphological preservation of parasites. However, if the organisms are very scanty, searching for them in a thin film may be very laborious. If mammalian blood is being studied, this difficulty can be partially overcome by making a thick film though the parasites in such films may be distorted and less well stained. Non-mammalian blood cannot be examined as a thick film since the erythrocyte nuclei interfere.

Thick blood films   Three or four drops of blood are collected on to the centre of a slide, and spread with a needle (or the corner of another slide) into a circular area about 1–1.5 cm in diameter. The film should be sufficiently thick for it to be just possible to read small print (e.g. the figures on a wrist-watch) through it. In order to make this thick preparation sufficiently transparent for microscopical examination, the blood, after thorough drying (about 12 h drying at room temperature, or 2–4 h in a desiccator at 37 °C, *not higher*, is desirable), is lysed by immersion in a hypotonic solution. This releases the contained hemoglobin from the erythrocytes, leaving any parasites more or less undamaged. It is, therefore, essential that thick films should not be fixed, since fixation prevents lysis. The staining procedure is as follows.

(1) Immerse the film in 0.5% aqueous methylene blue for 1 sec (this is not essential, but gives better results).
(2) Rinse the film briefly, *very gently*, either by dipping into water or holding under a *slowly* running tap.
(3) Place the slide face downwards in Giemsa's stain (prepared as described above for thin films) for 30 min (not longer).
(4) Continue as described for thin films (steps (3) and (4)).

Thick and thin films may be examined directly, or they may be mounted in a neutral mounting medium (such as "Euparal": Asco Laboratories) beneath a coverslip. If examined directly, a thin film of immersion oil should be spread gently over the slide before using nonimmersion objectives (oil is normally used with oil-immersion objectives, whether the film is mounted or not). For detection of protozoa, films are usually examined only with a × 100 oil-immersion objective, though preliminary scanning with a × 40 "high-dry" (i.e. nonimmersion) or (if available) × 50 oil-immersion lens may be used to locate larger forms (e.g. *Leucocytozoon*, some trypanosomes). Oil should be removed from unmounted films after examination by washing them in xylene which has been neutralized if necessary with calcium carbonate, or it may be removed with chloroform (*use xylene in a well-ventilated room and take care not to inhale the toxic vapor*).

Tissue impression smears

Small pieces of tissues such as lung, heart, liver, kidney, and spleen may be removed from dead animals. The cut surface is dabbed several times on filter paper or blotting-paper to remove most of the blood, and then pressed (not smeared) once on to a slide (several dabs may be made on one slide). After drying, such smears are fixed, stained, and examined in the way described above for thin blood films.

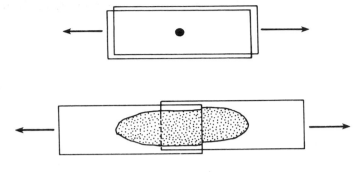

Figure 11.2 Diagram illustrating the preparation of a brain smear (see text for details).

Smears of brain, liver, and spleen may be made by placing a small piece (pin's head size) near one end of a slide, crushing it by placing another slide on it, and spreading it out by sliding the two slides longitudinally over one another (Fig. 11.2). Smears of bone marrow are usualy made by spreading a small fragment on a slide with the aid of a needle.

Histological sections

Viscera suspected of harboring parasites can be fixed, embedded, and sectioned by standard histological procedures. The sections may then be stained with hematoxylin and eosin or with Giemsa's stain. Small pieces of tissue for staining with Giemsa should be fixed for about 3 h in Carnoy's fluid (ethanol, 6 volumes: chloroform, 3 volumes: acetic acid, 1 volume) and washed in 90% ethanol before hydration, clearing, and embedding in the usual way. The staining procedure is given below.

(1) After hydration of the sections, place them in a mixture of 1 volume of Giemsa's stain, 1 volume of methanol and 10 volumes of distilled water at pH 7.2 (buffered as described above) for 1 h.
(2) Remove the slides from the stain and wash them in tap water.
(3) Differentiate the sections by pouring a 15% solution of colophonium resin in acetone on to the wet slides, rocking the latter to and fro for 3–10 sec, and pouring off the solution. Repeat this process two or three times until the blue-green dye no longer streams out in the differentiating solution.
(4) Wash the slides rapidly in a mixture of xylene (7 volumes) and acetone (3 volumes), followed by several washings in pure xylene (*beware of the toxic vapor*; use a fume cupboard if possible).

(5) Finally mount the slides in a *neutral* medium (e.g. "Euparal") beneath a coverslip.

Cultivation in vitro

Many species of *Trypanosoma* other than the Salivaria, *Leishmania* (and other Trypanosomatidae) can be grown in blood–agar media in test-tubes plugged with non-absorbent cotton wool, or screw-capped tubes or bottles. Inoculation of blood or pieces of tissue into such media can be used to diagnose infections which are too scanty to be detected by microscopical examination. Strictly sterile procedures must be used at all times, since bacteria and fungi grow readily in such media and trypanosomatids rarely develop in contaminated tubes.

Trypanosoma and *Leishmania* develop in cultures as the forms normally seen in the arthropod vector and not the vertebrate host (i.e. mainly as promastigote and epimastigote forms, though some species of *Trypanosoma* will develop metacyclic trypomastigote forms). Cultures must be kept at room temperature or 28 °C, not hotter (some species from fish require lower temperatures). To maintain the parasites in culture, a small drop of fluid from a flourishing culture is transferred aseptically to a tube of fresh medium once or twice a week (a Pasteur pipette, or bacteriological wire loop, is used to transfer the fluid). Details of one blood–agar medium are given below: there are many other equally suitable versions.

Dissolve any reliable proprietary nutrient agar (e.g. Difco, Oxoid) as recommended by the manufacturers in 1 litre of distilled water by steaming. While molten, dispense 5 cm^3 amounts into tubes or bottles and sterilize by autoclaving. When cooled to about 45 °C, add aseptically 1 cm^3 of blood (preferably previously inactivated by heating at 56 °C for 30 min, to destroy complement) to each tube, mix gently and place the tube at an angle of 40–50° so that the agar sets in a slope at the base of the tube. Horse, sheep, cattle, or rabbit blood is usually satisfactory; it may be obtained from commercial suppliers (e.g. Gibco, Flow Laboratories, or Oxoid). When the agar is set, add aseptically 1 cm^3 of any balanced, buffered physiological saline (e.g. Ringer's) to each tube. Penicillin, 200 units cm^{-3}, and streptomycin, 200 μg cm^{-3}, may be added to this saline.

Incubate the completed tubes at 37 °C for about 24 h to allow diffusion of nutrients into the saline, and also to check for contamination (if not sterile, signs of bacterial growth should be obvious by this time; the presence of any scum or precipitate, either on the agar slope or in the saline, suggests contamination and such tubes should not be used). Tubes may be stored at 4 °C for several weeks if desiccation is prevented by the use of screw caps or rubber caps.

Cultivation in vivo

Many protozoa can be grown in tissue cultures and in avian embryos, but such specialized techniques are seldom resorted to for purposes other than research. Certain parasites can also be maintained in suitable laboratory animals, but the facilities and techniques required are beyond the scope of this book. Similarly, the handling and dissection of invertebrate vectors of blood protozoa will not be dealt with here. Some sources of further information are listed at the end of this chapter.

Availability of material for study

The trapping and shooting of wild animals is, in many countries, restricted by law and is not, in any case, a thing to be undertaken lightly. Consequently, it is usually best to seek to obtain strains, or preserved material, of the tissue-dwelling protozoa from culture collections (e.g. The American Type Culture Collection), or other research or teaching laboratories (bearing in mind that such laboratories may not be able to deal with all requests), or from commercial biological suppliers if such are available. However, subject to the above provisos, blood films from any recently dead wild animals are often worthy of examination. Birds may be infected with *Plasmodium*, *Haemoproteus*, or *Leucocytozoon* in most countries; trypanosomes may also be present, but they are often too scanty to be detected readily. Wild rabbits, rodents, and insectivores may harbor trypanosomes, while the latter two groups may also be hosts to hemogregarines and piroplasms. These parasites, too, are cosmopolitan. Fish, the catching and killing of which are socially sanctioned, may be valuable sources of Microspora and Myxozoa in all countries of the world. The stickleback (*Gasterosteus aculeatus*) is commonly infected with a microsporidan, *Glugea anomala*. Fish, too, may be infected with trypanosomes, but parasitemias are often very low so that parasites may not be easily detectable on thin blood films.

FURTHER READING

Adams, K. M. G., J. Paul & V. Zaman 1971. *Medical and veterinary protozoology*, revised edn. Edinburgh: Churchill Livingstone.

Cable, R. M. 1970. *An illustrated laboratory manual of parasitology*. Minneapolis, Minnesota: Burgess.

Goldman, M. 1968. *Fluorescent antibody methods*. New York: Academic Press.

Jensen, J. B. (ed.) 1983. *In vitro cultivation of protozoan parasites*. Boca Raton, Florida: CRC Press.

Katz, M., D. D. Despommier & R. Gwadz 1982. *Parasitic diseases.* New York: Springer-Verlag.

Levine, N. D. 1973. *Protozoan parasites of domestic animals and of man*, 2nd edn. Minneapolis, Minnesota: Burgess.

Lillie, R. D. 1969. *Biological stains*, 8th edn. Baltimore, Maryland: Williams & Wilkins.

Melvin, D. M. & M. M. Brooke 1980. *Laboratory proceedings for diagnosis of intestinal parasites.* HHS publication no. CDC 80-8282. US Department of Health and Human Services. Atlanta, Georgia: Public Health Services, Centers for Disease Control.

Meyer, M. C. & L. R. Penner 1962. *Laboratory essentials of parasitology*, revised edn. Dubuque, Iowa, Wm. C. Brown.

Taylor, A. E. R & J. R. Baker (eds) 1985. *Methods of cultivating parasites* in vitro. London: Academic Press.

Taylor, A. E. R. & R. Muller (eds) 1971. *Isolation and maintenance of parasites* in vivo. Oxford: Blackwell Scientific.

ADDRESSES OF SUPPLIERS

This list is not meant to be exclusive; other brands may be equally satisfactory.

Allied Chemical and Dye Corporation, 40 Rector Street, New York 10006, USA.

American Type Culture Collection, 2301 Parklawn Drive, Rockville, Maryland 20852, USA.

Asco Laboratories, 52 Levenshulme Road, Gorton, Manchester, M18 7NN, England.

British Drug Houses (BDH Chemicals Ltd), Broom Road, Poole, Dorset, BH12 4NN, England.

Difco Laboratories, PO Box 1058-A, Detroit, Michigan 48323, USA.

Difco Laboratories, PO Box 14B, Central Avenue, East Molesey, Surrey, KT8 0SE, England.

Fisher Scientific Co., 711 Forbes Avenue, Pittsburgh, Pennsylvania 15219, USA.

Flow Laboratories Ltd, PO Box 17, Second Avenue Industrial Estate, Irvine, KA12 8NB, Scotland.

Gibco, 3174 Staley Road, Box 68, Grand Island, New York 14072, USA.

Gibco, Ltd, PO Box 35, Trident House, Renfrew Road, Paisley, PA3 4EF, Scotland.

Oxoid Ltd, Wade Road, Basingstoke, Hampshire, RG24 0PW, England.

Oxoid USA Inc., 9017 Red Branch Road, Columbia, Maryland 21045, USA.

Subject index

References to figures are in *italics*.

Index of generic and specific names

References to figures are in *italics*; references to tables are given under genera only, not species.